Grain Size Control

ALSO BY THE AUTHOR

The Physical Metallurgy of Microalloyed Steels

Grain Size Control

T. GLADMAN

Formerly Head of Physical Metallurgy, Swinden Laboratories, British Steel Corporation, and British Steel Professor of Metallurgy, University of Leeds, UK.

Routledge
Taylor & Francis Group
LONDON AND NEW YORK

FOR THE INSTITUTE OF MATERIALS, MINERALS AND MINING

First published for IoM^3 2004 by
Maney Publishing

2 Park Square, Milton Park, Abingdon, Oxfordshire OX14 4RN
605 Third Avenue, New York, NY 10017

Routledge is an imprint of the Taylor & Francis Group, an informa business

First issued in paperback 2020

. . .

Typeset by Old City Publishing, USA

ISBN 13: 978-1-904350-22-4 (pbk)

Contents

Preface		*ix*
Acknowledgemnts		*xiii*
Chapter 1	**The Nature of Grain Boundaries**	1
	1.1 Historical Views of Metallic Structures	1
	1.2 Modern Approaches to Polycrystalline Materials	4
	1.3 Grain Boundary Energy	6
	1.4 Grain Boundary Energy and Interfacial Tension	10
	1.5 Grain Boundary Forces	11
	1.6 Grain Boundary Segregation	15
	1.7 Summary	17
	References	18
Chapter 2	**Granular Arrays – Topology, Grain Shape, Stereology, and Grain Size**	21
	2.1 Topology	21
	2.1.1 General Principles	21
	2.1.2 Surface Energy and Space Filling Effects	23
	2.1.3 Implications for Grain Shape	24
	2.1.4 Experimental Assessment of Grain Shape	27
	2.2 Grain Size	29
	2.2.1 Stereological Principles	30
	2.2.2 Stereological Aspects of Various Polyhedra	35
	(a) Mean Areal Intersection	35
	(b) Mean Linear Intercept	37
	2.3 Experimental Measurement of Grain Size	40
	2.3.1 The Comparator Chart Method	40
	2.3.2 The Heyn Intercept Method	41
	2.3.3 Jeffries Planimetric Method	42
	2.3.4 Relationships Between Grain Size Measures	43
	References	44
	Chapter 2 Exercises	45
Chapter 3	**Grain Growth**	47
	3.1 Establishing a Granular Structure	47
	3.2 Normal Grain Growth	48
	3.2.1 Analytical Treatments	49
	3.2.2 Topological Models	56
	3.2.3 Computer Based Models	58
	3.2.4 Other Approaches	60
	3.3 Abnormal Grain Growth	61
	3.4 Growth Velocity and Impurity Effects	62
	References	65

Chapter 4 **Inhibition and Control of Grain Growth** 67
4.1 Introduction 67
4.2 Second Phase Particles 68
4.2.1 The Zener Relationship 69
4.2.2 Granular Arrays with Small Volume Fractions of Uniformly Sized Randomly Distributed, Second Phase Particles 71
(a) Growing Tetrakaidecahedra 71
(b) Shrinking Tetrahedra 74
4.2.3 Granular arrays with Small Volume Fractions of Uniformly Sized Grain Boundary Particles 76
(a) Growing Tetrakaidecahedra 77
(b) Shrinking Tetrahedra 78
4.2.4 Comparison of Grain Boundary Particles and Randomly Distributed Particles 79
4.2.5 The Effects of Mixed Particle Sizes 80
4.3 Free Surface Effects 84
4.3.1 Thermal Grooving 85
4.3.2 Variations in Free Surface Energy 88
4.3.3 Substrates and Capping of Foils 92
4.3.4 Secondary Recrystallisation in Foils or Thin Strip 94
4.4 Thermal and Mechanical Cycles 95
4.4.1 Thermal Cycling of Allotropic Materials 96
4.4.2 Thermomechanical Treatments 100
4.4.3 Conventional Controlled Rolling of Steel 102
References 108
Chapter 4 Exercises 109

Chapter 5 **Second Phase Particle Contents and Particle Coarsening** 111
5.1 Introduction 111
5.2 Particle Solubility 112
5.2.1 The Solubility Product 112
5.2.2 Using Simple Solubility Product 115
5.2.3 Calculating the Particle Contents for Simple Compounds 118
5.2.4 Solubility Data for Other Grain Refining Additives in Steel 119
5.2.5 Multiple Additions of Grain Refining Additives – Mutually Exclusive Compounds 123
5.2.6 Mutually Soluble Components 124
5.3 Conversions to Volume Fractions 129
5.4 Particle Size and Particle Coarsening 131
5.4.1 Precipitation 131
5.4.2 Ostwald Ripening 133
5.4.3 Implications of the Wagner Equation for Particle Selection 137
5.4.4 Calculation of Particle Growth Rates 140
References 140
Chapter 5 Exercises 141

Chapter 6 Applications in Grain Size Control 143
6.1 Steel Normalising Treatments 143
6.1.1 Fine Grained Normalised Steels 145
6.1.2 The Grain Coarsening Temperature 147
6.1.3 Secondary Recrystallisation 148
6.1.4 The Grain Size at Very High Temperatures in the Austenite Range 150
6.2 The Effects of Aluminium Content on Coarsening 151
6.2.1 The Solute Content 152
6.2.2 Effects of Prior Treatment 154
6.2.3 General Observations on Aluminium Treatment 157
6.3 Titanium Nitride Technology 158
6.4 Hot-rolled Products 164
6.4.1 Control of the Starting Grain Size 164
6.4.2 Inhibition of Grain Growth in Hot Working 166
6.4.3 Retarded Recrystallisation 166
6.5 Secondary Recrystallisation of Cold Worked and Annealed Strip 168
6.6 The Use of Oxide Particles 169
References 174

Answers to Excersies 177

Index 181

Preface

IMPORTANCE OF GRAIN STRUCTURES IN METALS

There are very few mechanical properties that are unaffected by either the grain size or the grain characteristics in polycrystalline metals. Two that readily spring to mind are the elastic modulus and the specific density. These properties depend on inter-atomic forces and the atomic structure of the particular metals. Most other properties show varying levels of sensitivity to grain size or grain orientations.

Probably the best known effect of grain size is that expressed by the Hall-Petch relationship, which states that the yield strength of a polycrystalline metal is related to the inverse of the square root of the grain size, and very significant increases in strength can be attained by maintaining a very small grain size. Similar effects are seen in other strength properties such as the ultimate tensile strength, and the fatigue strength.

The effect of grain size on the ductile cleavage transition temperature for b.c.c. metals is very pronounced. The transition temperature is negatively related to the inverse of the square root of the grain size, so that in order to avoid brittle cleavage fracture, the requisite low transition temperatures are attained by refining the grain size. The beneficial effect of fine grain sizes on the impact temperature makes grain refinement the only strengthening mechanism that will simultaneously improve toughness.

The effects of grain size on creep properties are more complex, though no less significant, and depend quite fundamentally on the creep conditions and the predominant creep mechanism involved. At relatively high stresses (or low temperatures) diffusion controlled dislocation creep predominates, and the creep rate is virtually independent of grain size. At relatively low stresses (or high temperatures) creep occurs by a bulk diffusion process involving transport of atoms from transverse boundaries to lengthways boundaries, and the creep rate is inversely related to the square of the grain size. Thus, smaller grains allow faster creep rates and reduce the creep strength. At intermediate stresses (or temperatures), creep occurs predominantly by a process of grain boundary diffusion known as Coble creep, and the creep rate is inversely related to the cube of the grain size. In order to give the extended lives required in creep service, stresses and temperatures are commonly set at levels in or near to the bulk diffusion regime and therefore fine grain sizes are not sought for creep resisting applications.

The grain size can also influence ductility at the high strain rates associated with metal forming if the deformation temperature is in a critical temperature range, due to the formation of cavities on transverse boundaries. At higher temperatures, rapid diffusion prevents cavitation, whilst at lower temperatures dislocation deformation predominates. This effect is also seen at lower temperatures when creep deformation rates are imposed. Intergranular cohesion can be reduced quite dramatically by the segregation of impurities to grain boundaries, and can result in brittle intergranular fracture. In material embrittled in this way, a ductile/intergranular transition temperature may be observed, the transition temperature increasing with increasing impurity concentration.

Other properties may depend on other grain characteristics such as grain orientation rather than grain size. The occurrence of a preponderance of grains of a certain orientation is described as a texture, and certain properties are known to depend on the type of texture produced. A classic example of this is the dependence of magnetic characteristics on the texture of thin strip for the high-silicon magnetic transformer steels. Development of a strong Goss texture can result in dramatic improvements of the efficiency of transformers. These textures have been developed traditionally by using a secondary recrystallisation treatment which encourages the abnormal growth of specific grains of favourable orientation. Another example of texture effects is to be found in sheet metal-forming of both ferrous and non-ferrous metals. Suitably textured material can be produced having superior sheet metal pressing properties, and also minimising the in-plane variability of strain that gives rise to earing in pressed sheet, which in turn reduces the waste from edge trimming.

There are literally thousands of publications illustrating the effects of grain size and grain structure on the properties of metals, but these are outside the scope of the present book. It should be clear from the above remarks, however, that the grain size can produce dramatic effects on the properties of metals, and indeed many alloy specifications call for specific levels of grain size control, or for property levels that can only be attained by such a control. The same can be said for texture control.

DEVELOPMENTS IN THE PAST FIFTY YEARS

It is almost fifty years since McLean (1957) published his book on 'Grain Boundaries in Metals'. This book provided an excellent account of the understanding of many matters concerning grains, grain boundaries, grain size, and associated phenomena existing at that time. Indeed, the understanding of grain structure at that time was very well advanced, in that many of the phenomenological features, associated with grains and grain/growth had been derived by the many physicists, chemists, and mathematicians, who had been attracted by the very catholic field of metallurgy, and which were soundly based on thermodynamic or other scientific argument. A particularly interesting case relates to the effects of small second phase particles in inhibiting grain growth. The effect of spherical particles had been derived by Zener, using classical surface energy physics for the interaction force and a simple collapsing bubble as the driving force for boundary motion, giving the classical Zener equation,

$$R = 4r/3f$$

where the grain radius, R, is related to the particle radius, r, and the volume fraction of particles, f. McLean rightly concluded that large inclusions in metals are probably unimportant in restricting grain growth, "but inclusions too small to be seen may limit the grain size, as, perhaps, in fine grained steels". This was indeed a prophetic statement, because at this time, the only solid fact was that additions of aluminium to steel would produce grain refinement and there was little indication of the nature of the growth inhibiting features, particularly as growth eventually occurred despite the continued presence of aluminium and it's stable compounds. It is important to realise that electron microscopy, invented in the early 1930's, was just starting to be used in studies of steel in the mid 1950's, thereby opening up the study of grain refining particles, especially when combined with electron diffraction and, at a later stage, energy dispersive and wavelength dispersive analytical techniques for these small particles. Also, theories of Ostwald ripening were not well developed. The original scientific concept of this phenomenon was available in the 1920's, but mathematical solutions to the coarsening problem were only developed later by Greenwood (1956) for a liquid matrix, and subsequently by Lifshitz and Slyozov (1961) and Wagner (1961) in connection with bulk diffusion in metal matrices. Yet another significant development that has occurred in this time interval is the introduction of computers of ever increasing cabability. It should be remembered that mechanical and electro-mechanical calculators were still being used in the early 1950's and most computers were of the valve type even in 1960. Whilst this does not in itself change the creative or critical capabilities of the human mind, it does provide an alternative approach to the investigation of grain growth phenomena, as will be shown later.

Thus, wide ranging developments have occurred since McLean published his book. Reference has been made previously to the developments in micro-structural examination techniques which include not only scanning and transmission electron microscopy, but also energy and wavelength dispersive X-ray analysis, and the various surface and interface spectroscopies such as ESCA and Auger spectroscopy, all of which have contributed to our understanding of the fine detail of metallic micro-structure. Reference has also been made to the introduction of computers, but due acknowledgement must be given to the hundreds (if not thousands) of workers who have each contributed many years of study to the subject of grain structure, grain control, and grain growth.

SCOPE OF THE BOOK

Despite the tremendous efforts devoted to these subjects, some problems remain unsolved, and in these cases, only the present state of understanding can be formulated together with an account of the remaining problems. In other areas, such as grain growth, there is an amazing degree of accord about the processes involved, and the qualitative aspects of micro-structural change, but an absence of a quantitative basis of prediction. In very few areas are the processes understood to an extent that quantitative agreement is obtained between experiment and theory. One area where this is possible is in the field of grain growth inhibition by second phase particles. One of the objectives of this book is to define the current state of understanding of these topics.

A second objective is to provide a guide to the significant literature. There are literally thousands of published papers each providing it's own contribution, but the sheer number of publications can be confusing to newcomers.

A further objective of this book is to provide a student text for undergraduates, postgraduates, and for continuing professional development. To this end, questions are included at the end of appropriate chapters, and solutions are given at the end of the book. It is hoped that these exercises will assist in the learning process, and will prove useful for the metallurgical (or materials) practitioner.

REFERENCES

Greenwood G. W., 1956. Acta Met. vol.4, p.243.

Lifshitz I.M. and Slyozov V.V., 1961. Journ Phys. Chem. Solids, vol.19, p.35.

McLean D., 1957. Grain Boundaries in Metals, Clarendon Press, Oxford.

Wagner C., 1961. Z. Electrochemie, vol.65, p.243.

Zener C., 1949. Private Communication to C. S. Smith, Trans. Amer. Min. Met. Engineers, vol.175, p.15.

Acknowledgements

The author would like to thank his many colleagues, co-workers, and students at British Steel and the University of Leeds for their help and assistance. Copyright permission from Student literature AB, Cambridge University Press, and Wiley VCH, is gratefully acknowledged. Other material from the literature is formally acknowledged in the references. Particular thanks are given to my wife, Dorothy, for her patience and contribution in the preparation of this book.

T.G

CHAPTER 1

The Nature of Grains and Grain Boundaries

1.1 HISTORICAL VIEWS OF METALLIC STRUCTURES

C.S. Smith (1951) has cited early studies of granular structures by Archimedes (3rd Century BC), Reamur (1724) and Lord Kelvin (1887). A sketch of the granular structure of wrought iron fractured in the hot short range is reproduced in Fig. 1.1, (Grignon, 1776). Studies of regular space filling shapes were carried out by Lord Kelvin (1887 and 1894). In 1887, Henry Clifton Sorby wrote "It seems nearly certain that the separate grains are separate, though imperfectly developed, crystals". This was based on his micro-structural studies of metals. Many practitioners believed that the structure of metals was amorphous because of the high ductility. Controversy continued for a further twenty-five years on whether metals were crystalline or amorphous, although evidence was building in favour of the crystalline state. Ewing and Rosenhain (1900), based on their studies of etch pits and slip lines, suggested that "the occurrence of such geometrical etch pits in the surfaces of metals that have never been polished may be taken as very strong evidence in support of the view that the crystalline grains of metals are built up of crystalline elements which are similarly oriented throughout the mass of each grain" – a very fair statement of the modern viewpoint. The foundations for solving the nature of metallic structures were finally laid by the X-ray diffraction discoveries of Von Laue and of Bragg in 1912.

The establishment of the theory of crystalline metals still left the problems of the nature of grain boundaries. Based on grain boundary sliding during elevated temperature deformation, Rosenhain and Humphrey (1913) postulated that the crystals of a pure metal are surrounded and cemented together by a very thin layer of the same metal in the amorphous condition. This was known as the amorphous cement theory. Subsequently this was replaced by the transition lattice theory in which very few rows of atoms (one or two) occupied sites that were a compromise between those required by the two lattices of the adjacent grains. This had been suggested as a possible grain boundary structure by Jeffries and Archer (1924), and was later described unambiguously by Hargreaves and Hills (1929), who pointed out that according to the transition lattice theory the energy and properties of grain boundaries would depend on the misorientation of the lattices on either side of the boundary, whereas, according to the amorphous cement theory, all boundaries should be the same, irrespective of the lattice orientations. Despite the observation made by Chalmers (1937) that the obstruction offered to slip along the glide plane depended upon the lattice orientations on either size of the boundary, suggesting a transition lattice, grain boundary sliding rates at elevated temperatures were still difficult to reconcile with this

Fig. 1.1 Sketch of an Aggregate of Iron Grains by Grignon in 1776. (After Smith, 1952).

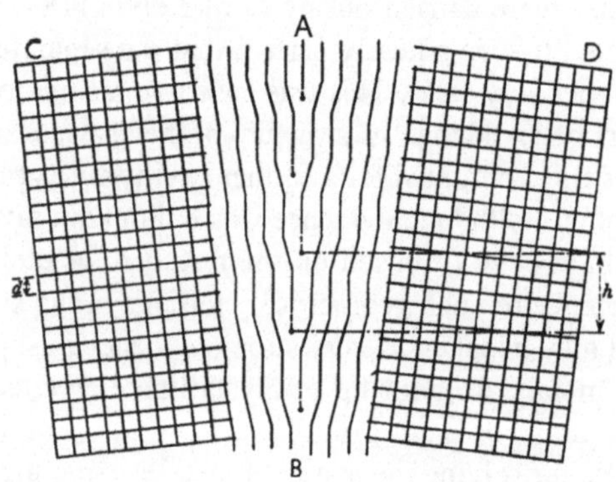

Fig. 1.2 A Wall of Edge Dislocations Causes a Sharp Change of Lattice Orientation.

structure. Mott (1948), with his 'island' model, and Ke (1949) with his diffusion model, attempted to explain how thin (transition type) boundaries only one or two atoms thick could permit extensive intergranular sliding at high temperatures. Although these were partially successful, the predicted rates of sliding were still two or three orders of magnitude too low. A feature of both of these models was the concept of areas of the grain boundaries having good atomic matching and other areas showing poor atomic matching. Grain boundary sliding in the easier of these areas (there was some disagreement here) enhanced the stress on contiguous unslipped area, causing the slipped regions to expand.

It was at about this time that the dislocation theories of grain boundaries were being developed, culminating in the signal work of Read and Shockley (1950). These are essentially transition lattice theories, suggesting a narrow grain boundary with serious atomic displacements being confined to about 3 atomic spacings. The concepts of tilt and twist boundaries were put forward and it was suggested that any orientation differences between pairs of grains could be accommodated by judicious positioning of numbers of edge and screw dislocations. Consider the case of a simple symmetrical tilt boundary, Fig. 1.2. At low tilt angles, the misorientation can be accommodated by a row of edge

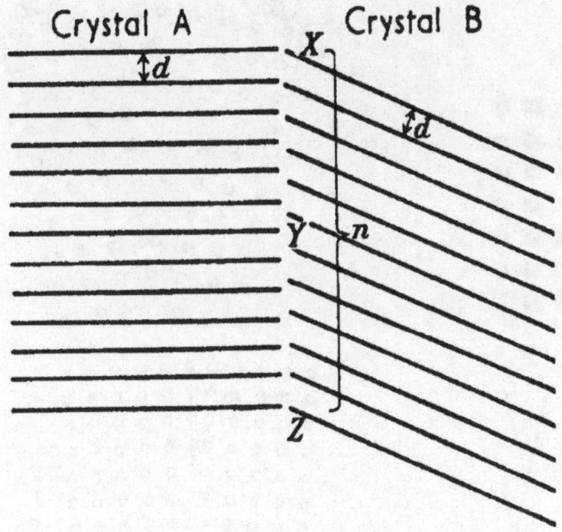

Fig. 1.3 A Low Angle Boundary Showing the Periodicity of the Mis-match.

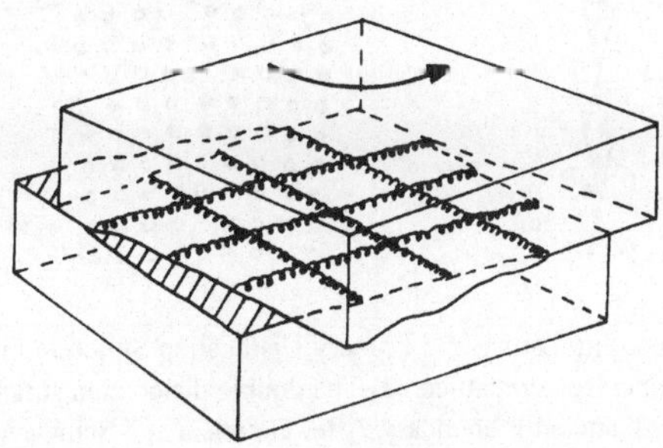

Fig. 1.4 Twist Boundary Produced by a Grid of Screw Dislocations.

dislocations, and the spacing 'h' between these dislocations will decrease as the angle of tilt between the two grains increases. Burgers (1953) and Bragg (1952) had in fact pointed out that a wall of dislocations such as this would give a relatively sharp tilt to the lattice on either side. In the case of an asymmetric tilt boundary, Fig. 1.3, there is a mismatch in the atomic spacings on either side of the boundary, but because of the periodicity of the lattices, there is a pattern of coincident atom positions, the mismatch first increasing on moving from a coincident point and then decreasing to the next coincident point, i.e. a periodicity of the stress pattern. Whilst relatively small mismatches can be accommodated elastically, extended regions of high mismatch would have a high elastic energy level. This can be reduced by the formation of a grain boundary dislocation.

Rotation of one crystal relative to another can be accommodated by the introduction of one or more sets of screw dislocations, Fig. 1.4. Again, increasing the twist angle decreases the dislocation spacing in the twist boundary. Both the tilt and twist boundaries described are of a particularly simple kind allowing relatively small differences in orientation. Usually, grain boundaries separate crystals that have large angle misorientations. Grain boundaries have five degrees of freedom, three from rotations about the three perpendicular axes of one lattice with respect to the other and two from the two tilts of the planar grain boundary surface with respect to either of the crystal lattices. Nevertheless, any grain boundary may be described by an arrangement of edge and/or screw dislocations,

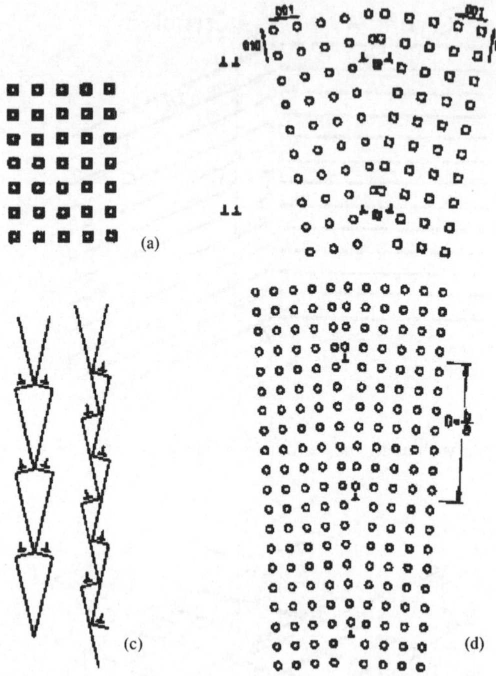

Fig. 1.5 Correlation of Coincident Site Lattice and Primary Dislocation Structure in a Grain Boundary. (a) Two identical lattices. (b) Coincident site lattice. (c) The double dislocation structure relaxes to give – (d) Dislocation structure of a small angle tilt boundary. After Gottstein and Schwarzer.

but, in the case of high angle boundaries, their arrangement will be complex, so that individual dislocations are not easily recognised. These boundaries will have islands of good fit, interspersed with areas of bad fit, where the bad fit is accommodated by the introduction of intrinsic secondary grain boundary dislocations, which are not necessarily lattice dislocations, but which are necessary to maintain the boundary misorientation. These secondary grain boundary dislocations are necessary at large misorientations because any attempt to achieve high tilts or high twist boundaries using lattice dislocations would require so many dislocations, packed so closely together, that the dislocation cores would overlap. This is energetically unacceptable.

1.2 MODERN APPROACHES TO POLYCRYSTALLINE MATERIALS

Grain boundary structures have attracted considerable interest and effort over the past forty years, and we now understand more about the periodicity of grain boundaries. One approach to the problem of grain boundaries was to consider the atom positions in the two lattices across the grain boundary. Fig. 1.5, after Gottstein and Schwarzer (1992), shows the development of a low angle tilt boundary. If two identical cubic lattices with parallel axes (a) are symmetrically rotated about an [001] axis normal to the plane of the diagram, coincident sites occur at regular intervals along the tilt boundary, as in (b). The

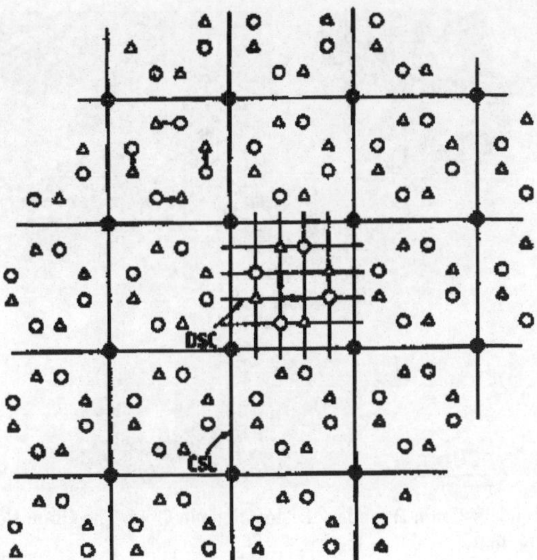

Fig. 1.6 Coincident Site Lattice and the Displacement Shift Complete Lattice for a Twist Boundary. (Gottstein and Schwarzer).

corresponding double dislocation arrangement is relaxed by translation along the boundary, as in (c), yielding the dislocation structure of a small angle symmetrical tilt boundary, as in (d). However, the concept of the coincident site lattice (CSL) is not well illustrated for the tilt boundary, as the translation process effectively destroys the coincident lattice sites, and presents a relatively sharp lattice curvature across the boundary. Nevertheless, the periodicity of stress from one edge dislocation to the next is clearly illustrated by the increased separation of the atoms below the edge and the compression of atoms above the edge as is seen for any edge dislocation.

The concept of the CSL is probably best illustrated by considering a twist boundary as shown in Fig. 1.6. In this case, the two cubic lattices are mutually rotated about an [001] axis normal to the plane of the diagram, but the grain boundary now lies in the plane of the diagram. The type of twist boundary is usually described by Σ, the reciprocal of the fraction of atoms in the boundary plane that occupy coincident sites. For the example shown, the lattice rotation of 36.9° produces a coincident site lattice in which one fifth of the atoms lie on the CSL, i.e. Σ=5. The basic idea behind the coincident lattice site theory was that, with such a large proportion of atoms in registration across the boundary, the boundary should have a low energy. However, it is clear from Fig. 1.6 that the remaining four fifths of the atoms show considerable disregistry across the boundary, and the displacement required for each fifth of these atoms is in a different direction, as shown towards the upper left corner of Fig. 1.6. If one of the lattices is translated to bring one of the disregistered atoms into registry, a new but identical coincident site lattice is created. This in turn has lead to the concept of the displacement shift complete (DSC) lattice, shown near the centre of Fig. 1.6.

Any slight deviation from exact coincidence orientation relationship will destroy the CSL, and the associated periodicity. It is possible however to preserve the periodicity by introducing secondary grain boundary dislocations (SGBD), which localise the mismatch but conserve the long range order just as lattice dislocations do in low angle boundaries.

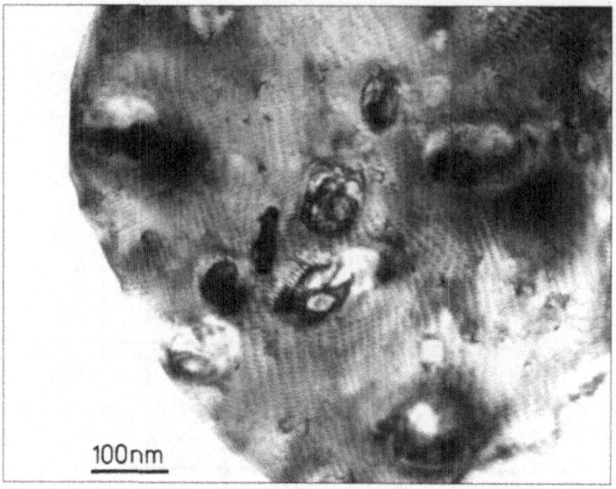

Fig. 1.7 Networks of Secondary Grain Boundary Dislocations in Gold. The Grain Boundary has a mis-orientation near to Σ5 (after P. Goodhew).

The Burgers vector of the SGBDs can be much less than that of the lattice translation dislocations, and clearly must be vectors of the DSC lattice. There is now experimental evidence for the existence of SGBDs, Fig. 1.7, for orientation relationships that are slightly off the coincidence rotations.

Contrary to expectations, measured grain boundary energies do not generally show significant reductions at high coincidence levels (low Σ) except for the Σ=3 (coherent twin) boundary and Σ=11 symmetrical tilt boundaries. Consequently, atoms at the boundary have been allowed to adjust to minimise short range effects of co-ordination and potential with respect to near neighbours, rather than preserving either lattice geometry. This has been done by computer methods and will be explored in greater detail in the next section.

1.3 GRAIN BOUNDARY ENERGY

The small angle tilt or twist boundaries containing uniformly disposed lattice dislocations may have an energy that can be calculated from a knowledge of the dislocation density and the line energy. Read and Shockley (1950), showed that the grain boundary energy, γ, is obtained from the energy of the periodic dislocation arrangements, by the following equation

$$\gamma = E_0\theta(A - \ln\theta) \qquad \text{(equ 1.1)}$$

Where $E_0 = Gb/4\pi(1-\nu)$

$A = 4\pi(1-\nu)\, B_e/Gb^2$

b = Burgers vector

G = shear modulus

ν = Poisson ratio

B_e = energy in the dislocation core

θ = angle of misorientation (radians)

It should be noted that, in the absence of information on the core energy, an absolute comparison with experiment is not possible. However, differentiating equation 1.1 gives

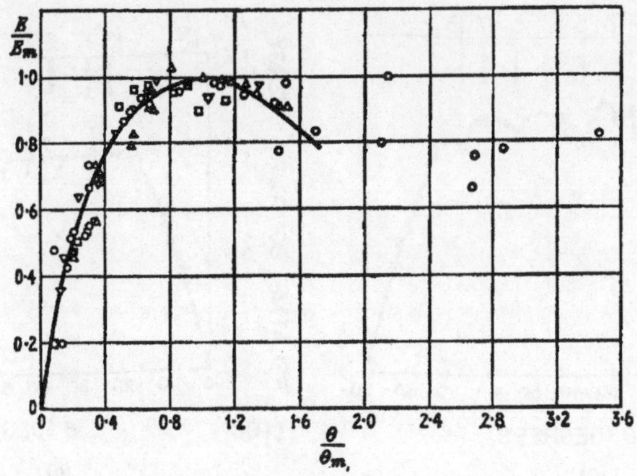

Fig. 1.8 Tile Grain Boundary Energy as a Function of Mis-orientation. Data points show values for silicon iron, tin, and lead.

$$d\gamma/d\theta = E_0(A - \ln\theta) - E_0 \qquad \text{(equ. 1.2)}$$

and, when γ is a maximum, $(= \gamma_m)$, $d\gamma/d\theta$ is zero, so that, from equation 1.2

$$A - \ln\theta = 1 \qquad \text{(equ. 1.3)}$$

where θ_m is the misorientation (in radians) at which the boundary energy is at a maximum. Substitution from equation 1.3 into equation 1.1 gives

$$\gamma_m = E_0\theta_m \qquad \text{(equ. 1.4)}$$

Dividing equation 1.1 by 1.4 and further substituting for A, yields

$$\gamma/\gamma_m = (1 - \ln\theta/\theta_m)\,\theta/\theta_m \qquad \text{(equ. 1.5)}$$

This expression compares well with the experimental work of Dunn and Lionetti (1949), and Aust and Chalmers (1950 and 1951), as shown in Fig. 1.8. Even here, however, it is necessary to select a value of θ_m that gives best fit with the experimental data points. One of the conditions of the derivation of equation 1.5 is that the misorientations should be small so that they are accommodated by lattice dislocations, and also that the approximation $\theta = \sin\theta$ is valid. The fit between experimental data and theory appears to show reasonable agreement up to values of θ/θ_m of 1.5, which, with values of θ_m up to 29°, suggests that the data is valid for angles of θ up to about 45°, which can hardly be described as low angle boundaries. Also, with the very high values of θ/θ_m (from 1.7 to 3.5) one might have expected to see low values of grain boundary energy in the experimental data, due to the mirror image of tilt angles above and below the 45° tilt in the cubic metals, as suggested by Van de Merwe, (1950).

More recent computer based calculations of the grain boundary energy for [100] and [110] tilt boundaries in aluminium, have suggested that very low energies may be associated with low value coincident site lattices. Indeed some measurements have revealed low energy boundaries at certain of these positions as shown in Fig. 1.9,(Hasson and Goux, 1971). This is particularly true for the $\Sigma = 3$ coherent twin boundary. The measured grain boundary energies for the <100> and <110> tilt boundaries show the same qualitative dependence on the tilt angle as the computed values. It is interesting to note that

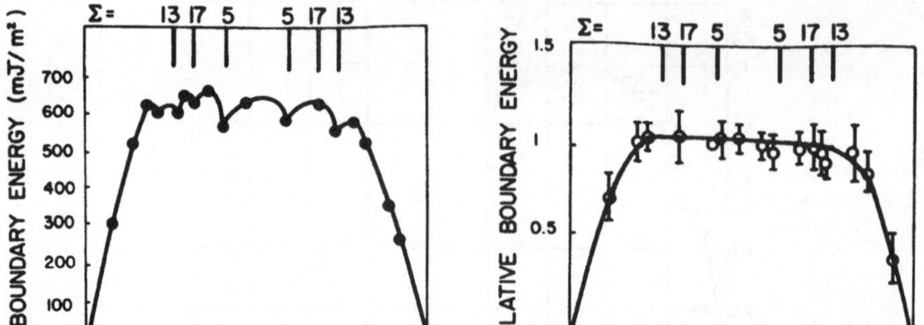

Fig. 1.9 Computed and measured Energies for Symmetrical <100> and <110> Tile Boundaries in Aluminium. (a) and (c) are computed.

there aren't any major cusps (low energy positions) for the [100] tilt boundaries at low values of coincident lattice sites, e.g. Σ = 5. Gottstein and Schwarzer (1992) point out that atomic relaxation processes made to minimise the lattice distortion energy associated with the boundary will also tend to displace the coincident site atoms, thereby destroying the CSL. For example, in Fig. 1.6, if the non-coincident atoms within the coincident site lattice rearrange themselves to give better fit, they will produce a knock on effect for atoms at the coincidence sites, thereby removing the coincidence lattice.

A further point made by Gottstein and Schwarzer (1992) is that the calculated energies are absolute energies, and do not necessarily represent the free energy that is minimised by the atomic rearrangement. Derivations of the free energy from the calculated energy values involves entropy, and a temperature dependence of grain boundary energy is immediately implied. Direct evidence for temperature effects on the grain boundary energy have been demonstrated by Erb and Gleiter (1979), and by Shvindlerman and Strumal (1985). Equally as strong as the effect of temperature, is the effect of impurity elements. Many soluble impurities are known to segregate to grain boundaries, see section 1.6. This effect would be expected to lower the energy of boundaries having greatest misfit, thereby causing less

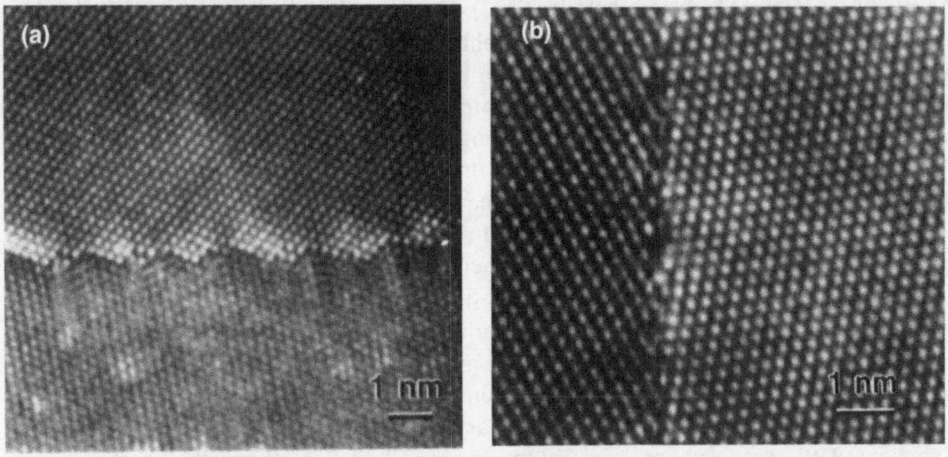

Fig. 1.10 High Resolution Electron Microscopy. A single grain boundary can dissociate into symmetric and asymmetric micro-facets.

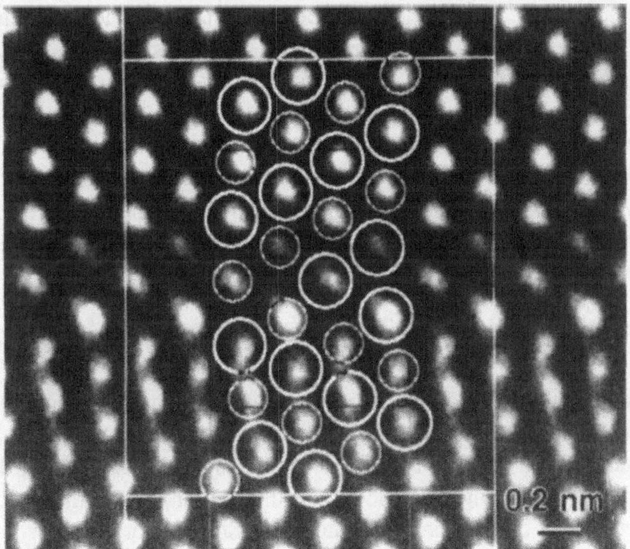

Fig. 1.11 Digitally Averaged Structural Units of a (ll3) (113) Grain Boundary. The circles indicate the position of atomic columns in the HREM image. (After Merkle and Wolf).

variation between grain boundary energies, (Santer, Gleiter and Baro, 1977). An interesting idea was put forward by Goodhew (1979), suggesting that boundaries may dissociate to give a new grain of such an orientation that it would form two low energy boundaries with neighbouring grains. However, in the example cited, one of the boundaries reportedly formed by dissociation was a coherent twin boundary of $\Sigma = 3$, and the question must be raised as to whether all coherent twins are formed in this way. Direct experimental support of boundary disintegration, however, has been obtained by high resolution electron microscopy, (HREM). The high resolution micrograph in Fig. 1.10(a), shows a well facetted boundary, (Merkle and Wolf, 1992). The boundary is characterised by having segments showing two different crystallographic planes on the adjacent crystals. In the case of Fig. 1.10(a), the general (225)(441) grain boundary has disintegrated into (111)(001) asymmetric micro-facets. Even more complex dissociations are observed such as that shown in Fig. 1.10(b). Here, the HREM micrograph of a <110> tilt grain boundary in gold (tilt angle of 39°) shows a (221)(221) symmetric boundary facet near the top, changing to an asymmetric (557)(771) facet in the lower part of the picture.

High resolution electron microscopy has further provided direct evidence of the nature of grain boundaries, Fig. 1.11. This shows the structure of a (113)(113) tilt boundary. The superimposed circles indicate the assumed positions of the atomic columns constituting the grain boundary. Based on such detailed information about the boundary, grain boundary energies have been correlated with volume changes, co-ordination coefficient, and the number of broken bonds per unit area of grain boundary. In fact, the concept of a volume expansion dependence of the grain boundary energy was put forward by Seeger and Schotky (1959) over forty year ago, but the very small dilations (typically less than a very small fraction of the lattice spacing) has made measurements impossible until recently.

Most of the work carried out on boundary structure has considered relatively simple tilt and twist boundaries, and despite the presence of special coincidence boundaries with low Σ values, very few low energy boundaries have been found experimentally. Bearing in mind that symmetrical boundaries can be formed by tilting around a large number of rotational axes, that there will be an equally large number of asymmetric boundaries, and that these will be supplemented by a vast number of twist boundaries covering a wide range of rotational axes, then the occurrence of high coincidence boundaries (low Σ) having low grain boundary energies will be rare. It would seem reasonable, therefore, to assume that grain boundaries in general have similar high energies, with very few low energy boundaries, the most notable one being the coherent twin boundary.

1.4 GRAIN BOUNDARY ENERGY AND INTERFACIAL TENSION

Given the computed and observed structures of grain boundaries, the energy of atoms in the boundary position is higher than those in the grain interior, and experiments and calculations also confirm that the free energy of boundary atoms is higher that that of the grain interior atoms, since prolonged heating will allow adjustments that reduce the total boundary area, resulting in a net transfer of grain boundary atoms to grain interior positions.

Grain boundaries also possess an interfacial tension. The interference micrograph shown in Fig. 1.12 was obtained from a thermally etched copper-antimony alloy, McLean (1957). The grain boundary tension is balanced against surface tension to give the characteristic grooves where the grain boundaries intersect the surface. The balance of surface tension with grain boundary tension is also reflected in the 'line tension' effected, noted by McLean, where there are more interference fringes seen at the two triple junctions.

In the case of a liquid, it is well known that the surface tension is numerically equal to the surface free energy when both are expressed in the same units, such as dynes per centimetre and ergs per square centimetre. This is deduced by a virtual work argument. If the surface tension, T, acts over a length of interface, w, the total force, F, is wT. If this would cause the length of interface to advance by an amount dl, then the work done would be F.dl or Tw.dl. The advance of the length by dl would increase the area by w.dl. If the surface free energy is γ per unit area, then the change in free energy is γw.dl, and equating this with the work done yields $\gamma = T$. This

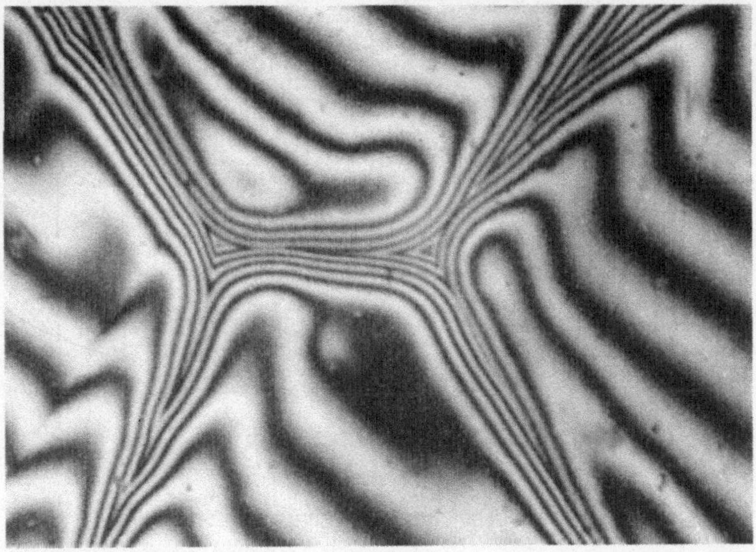

Fig. 1.12 Interference Micrograph of a Copper/Antimony Alloy. Line tension at the two triple junctions has pulled in the surface and caused the extra fringes. x 1000.

may not work for crystalline solids as the surface energy may depend upon the crystal orientation with respect to the surface, and the elastic anisotropy may also have an effect.

However, any possible numerical inequality does not preclude certain parallels being drawn between surface tension and surface energy effects. Any movement of the boundary which results in a reduction in area of the boundary, and therefore a reduction in energy, can be regarded as causing a change in energy for each unit of distance moved – the classical definition of a force. The magnitude of the force is described by the rate of change of energy per unit of advancement.

1.5 GRAIN BOUNDARY FORCES

A direct consequence of the grain boundary energy, and hence the driving forces for grain boundary movements can be seen in the nature of grain boundary junctions. Consider a four-rayed junction such as is shown in Fig. 1.13. Although this is a two dimensional diagram, it can be considered as a plane section normal to the grain boundary junction, and if a second parallel plane section is made at a unit distance from the first, the lengths of the segments will be numerically equal to the areas of the segments between the two planes. The boundary lengths between the four grains shown in Fig. 1.13 can be reduced by allowing the four-rayed grain boundary junction to degenerate and become two three-rayed junctions, so that the grains A and B now meet to yield a segment PQ, rather than meeting at the point O. As the points P and Q move apart, the total length of the segments AP, BP, CQ, DQ and PQ initially decreases, and reaches a minimum, before eventually increasing. This minimum length has been shown (Gladman 1997) to coincide with the triple point angles reaching 120°. If it is assumed that all five grain boundaries have the same energy per unit area, and that the grain boundary energies are unaffected by rotation of the

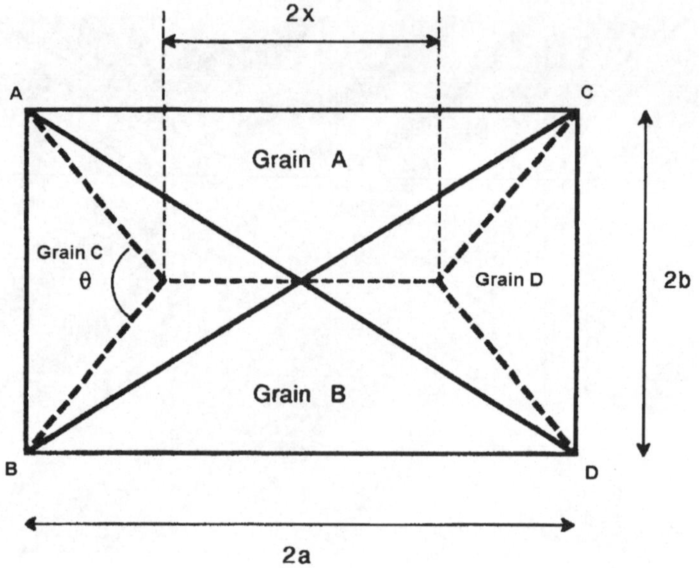

Fig. 1.13 The Four-rayed Junction at O Dissociates into Two Triple Junctions at P and Q to Lower the Boundary Energy.

segments, AP, BP, CQ and DQ, then the minimum in the lengths of the segments will correspond with a minimum in grain boundary energy. Thus, a four-rayed junction, i.e. an edge where four grains meet, will be unstable, and will degenerate to become two three-rayed junctions, with three grains meeting at each of the two grain boundary edges (i.e. the triple points in Fig. 1.13).

The mathematical solution appears to be independent of the dimensions of the initial framework represented by the fixed nodes of the rectangle ABCD, but a more critical examination shows that if the ‘a’ dimension were substantially less than the ‘b’ dimension, then the new segment of grain boundary would be vertical in Fig. 1.13 and grains A and B would be separated. This would not happen in the specific case shown in Fig. 1.13, because the angles AOC and BOD already exceed 120°, so that degeneration of the four-rayed junction to give a segment of boundary between grains C and D would further increase these angles and increase the grain boundary energy.

In order to examine the effects of the fixed nodal points of the rectangle, consider the lengths of the segments as a function of the displacement x. Because of the symmetry of Fig. 1.13, only the right hand half of the rectangle need be considered. The total length, L, of the segments CQ, DQ and OQ is given by

$$L = 2(b^2 + (a^2 - x^2))^{1/2} + x \qquad \text{(equ. 1.6)}$$

The rate of change of L with x is given by

$$dL/dx = 1 + 2(x\text{-}a)/(b^2 + (a - x)^2)^{1/2} \qquad \text{(equ. 1.7)}$$

In order to reduce the segment length then dL/dx must be negative, i.e. correspond with a reduction in energy, at very small values of displacement, x. dL/dx at x = 0 is given by

$$dL/dx = 1 - 2\,a/(b^2 + a^2)^{1/2} \qquad \text{(equ. 1.8)}$$

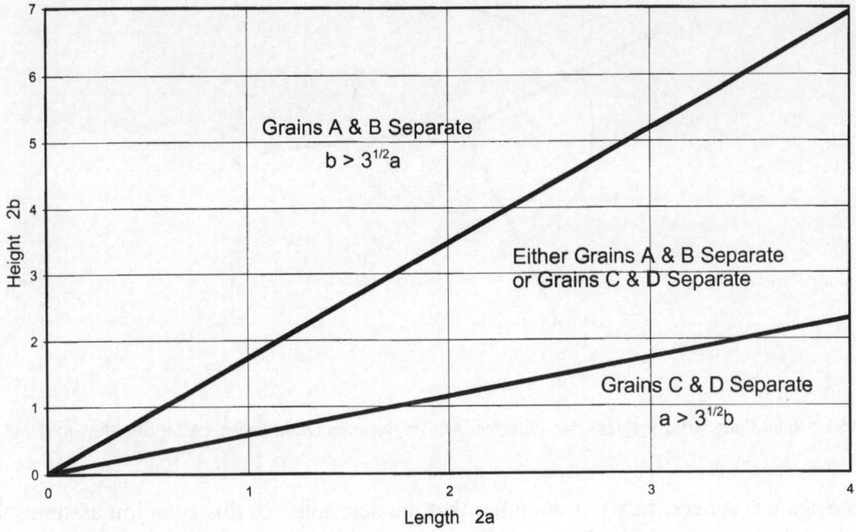

Fig 1.14 Effect of Frame Shape on Movement of Triple Points.

which is negative when

$$2a/(b^2 + a^2)^{1/2} > 1 \qquad \text{(equ.1.9)}$$

which in turn reduces to $3^{1/2}a > b$. It can be seen that the displacement, x, of the triple point can lead to a reduction in the line lengths (energy of the system) even when b is larger than 'a', but not when b is larger than $\sqrt{3}a$. Because of the model symmetry, it also follows that a vertical boundary between grains can be formed when 'a' is less than $\sqrt{3}b$. As shown in Figure 1.14, this leads to a zone where either a horizontal or a vertical segment may form i.e. between $b = \sqrt{3}a$ and $b = a/\sqrt{3}$. Outside this range the central boundary will only form parallel with the long axis. Within this range, either a vertical or a horizontal boundary could form, but it is likely that the boundary parallel to the longer axis would form as the rate of change of energy is greater when the more acute angle is broadened. However, irrespective of the positioning of the fixed nodal points, the four-rayed junction would decompose into two three-rayed junctions, and the lowest energy position would occur when the angles of the triple points reach 120°. It should be remembered that this conclusion is only valid if it is assumed that grain boundaries have the same energies, irrespective of either the lattice orientation or the boundary orientation. These may seem to be sweeping assumptions but grain boundary energy measurements, and observed equilibrium structures, suggest that only for very special boundaries is the grain boundary energy significantly reduced.

The more general case of triple points, with different grain boundary energies has been investigated (Shuttleworth, 1952). The equilibrium positions of the three grain boundaries are as shown in Fig. 1.15. The equilibrium positions are defined by

$$\gamma_1/\sin\alpha_1 = \gamma_2/\sin\alpha_2 = \gamma_3/\sin\alpha_3 \qquad \text{(equ 1.10)}$$

where γ_1, γ_2, and γ_3 are the grain boundary energies

and α_1, α_2 and α_3 are the corresponding opposite angles.

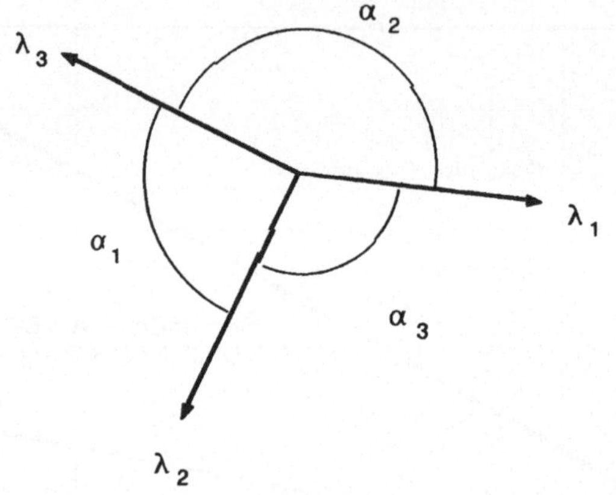

Fig. 1.15 Unequal Boundary Energies Give Different Angles Between Grain Faces. (After Shuttleworth).

Once again it is necessary to remember that the derivation of this equation assumes that even though the grain boundary energies may be different, the boundary energy is unchanged by rotation. A detailed consideration of such effects concluded that any corrections are generally small, and only in special cases, e.g. a twin boundary, would the boundary orientation cause significant changes in grain boundary energy.

Experimental support for the generally similar levels of grain boundary energy was first noted (Desch, 1919), simply by observing that most triple points in a metallograph have near 120° angles. Support also came from an examination of polyphase alloys (Smith, 1948). $\alpha - \alpha$ boundary triple points showed near 120° angles, i.e. near constant grain boundary energies, but that $\alpha - \beta$ boundaries showed a consistently different energy from that of the $\alpha - \alpha$ boundaries, the $\alpha - \beta$ boundary energies again being almost constant. The arrangement at triple points involving two a – grains and one β grain, gave a characteristic arrangement of angles as shown in Fig. 1.16, the triple junction again being normal to the plane of the diagram. Smith used the term "dihedral angle" to describe the angle between the $\alpha - \beta$ boundaries. Resolving the tensions parallel to $\gamma_{\alpha-\alpha}$ in fig. 1.16

$$\gamma_{\alpha-\alpha} = 2\,\gamma_{\alpha-\beta} \cdot \cos\theta/2 \qquad \text{(equ 1.11)}$$

Studies of duplex $\alpha - \beta$ brasses also included a leaded brass, the lead being added to improve machinability. The dihedral angle for the β phase against $\alpha - \alpha$ boundaries was found to be 95°. From equation 1.11,

$$\gamma_{\alpha\beta} = 0.74\gamma_{\alpha\alpha} \qquad \text{(equ. 1.12)}$$

The dihedral angle at a triple point for the lead phase against α - α boundaries was found to be 90°, yielding –

$$\gamma_{\alpha\text{-Pb}} = 0.707\,\gamma_{\alpha\alpha} \qquad \text{(equ. 1.13)}$$

Finally the three angles at a triple point between the α-phase, the β-phase, and lead were observed to be 110° opposite the α-β boundary, 120° opposite the α-Pb boundary and 130° opposite the β-Pb boundary. From Equation 1.10 –

$$\gamma_{\alpha\beta}/\sin 110° = \gamma_{\alpha Pb}/\sin 120° \qquad \text{(equ. 1.14)}$$

$$\text{or } \gamma_{\alpha-\beta}/\gamma_{\alpha\text{-Pb}} = 1.085$$

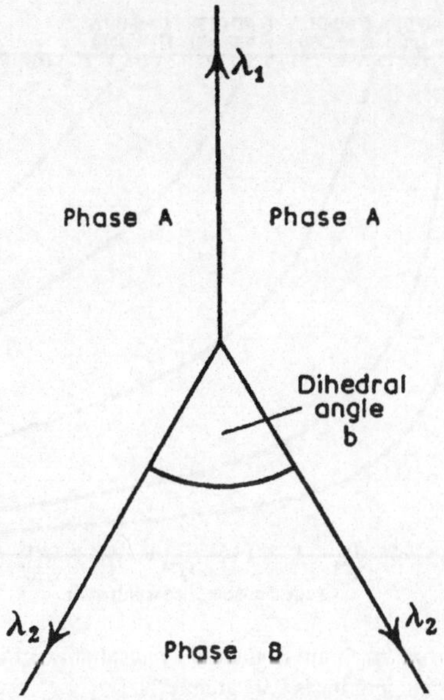

Fig. 1.16 (right) The Dihedral Angle at the Triiple Junction between a B-phase Grain and Two A-phase Grains.

Using the results found from equations 1.12 and 1.13 gives an independent estimate of $\gamma_{\alpha-\beta}/\gamma_{\alpha\text{-Pb}}$ of 0.74/0.707 = 1.047, which is reasonably close to that derived from equation 1.14. It should be pointed out that the angle seen in the micrograph may not be the dihedral angle, due to random variations between dihedral angles and the plane of section. Studies of the statistical distribution arising from the random sectioning (Smith 1948, Harker and Parker, 1945) showed that the most frequently observed angle would be the true dihedral angle. An important feature of this type of study is that the grain boundary energy does appear to be a general characteristic of the material, and although there are notable exceptions, the consistency of relative grain boundary energy measurements suggests that these are few.

1.6 GRAIN BOUNDARY SEGREGATION

The current view of a grain boundary is that the transition region from the lattice of one grain to that of the opposite grain occurs over a short distance of only two to four atom rows, i.e. the boundary is thin. Within the boundary, there may possibly be regions of relatively good fit (or correspondence) between the two lattices, separated by areas of greater misfit, possibly due to the presence of lattice dislocations or secondary grain boundary dislocations. The resulting effect is, overall, that the atoms are somewhat less densely packed in some part of the boundary than they are in the grain interiors, but there is some variation in the packing, so that other regions have more tightly packed atoms.

One of the consequences of the grain boundary structures is that certain solutes which do not fit easily into the atomic lattice of the host atoms, may fit more easily into certain of the grain boundary sites. These are termed horophile impurities. In steels, elements from groups IVB and VB, such as phosphorus, arsenic, antimony and tin, are known to segregate strongly to grain boundaries. The thermodynamically based relationship between the equilibrium

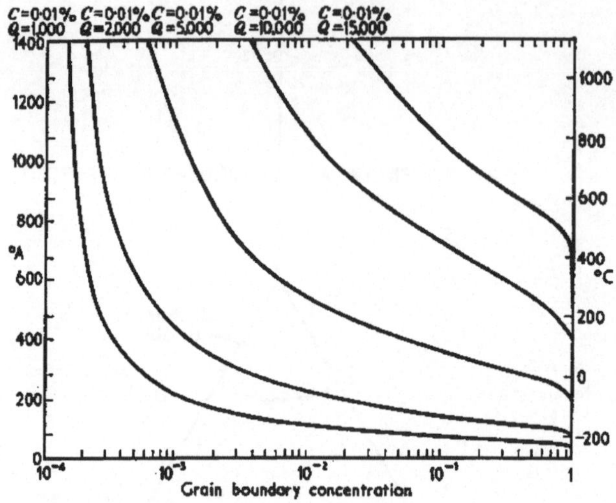

Fig. 1.17 The Effect of Temperature on the Grain Boundary Concentration of Impurities of Different Binding Energies. The concentration in the grain interior is 0.01 atomic %.

grain boundary concentration and the bulk concentration is described (Slater, 1939) as

$$C_b = C_o \exp(Q/R\,T)/(1 + C_o \exp(Q/RT))$$

Where C_b is the boundary concentration

C_o is the bulk concentration

Q is the binding energy between solute and grain boundary sites

R is the gas constant

andT is the absolute temperature.

The importance of this relationship is that even with very low bulk solute contents, and relatively modest binding energies, the grain boundary concentration of the solute can be high, as indicated in Fig. 1.17. With low binding energies, very high grain boundary concentrations would be expected only at very low temperatures, and it is doubtful whether the kinetics of diffusion would allow equilibrium concentrations to be attained in practicable times. Binding energies in the range 40 – 60 J/mole/K would however give rise to high grain boundary concentrations of solute at temperatures high enough to make the kinetics favourable. A further important feature of Fig. 1.17 is that the solute content at the grain boundaries decreases dramatically at higher temperatures.

The propensities of a range of solutes to segregate to grain boundaries in iron, copper, nickel, gold, and tin alloys, are shown in Fig. 1.18 where the grain boundary enrichment factor, C_b/C_o, is shown as a function of the limit of solid solubility of the solute, both being measured at the same temperature for each base and solute. As the solid solubility decreases, so the partition coefficient increases (Hondros and Gupta, 1979; English and Gladman 1986).

The significance of this phenomenon in relation to grain growth is that any grain boundary movement which is not accompanied by a corresponding movement of the grain boundary solute, will require a large additional energy to detach the boundary from the solute. In the

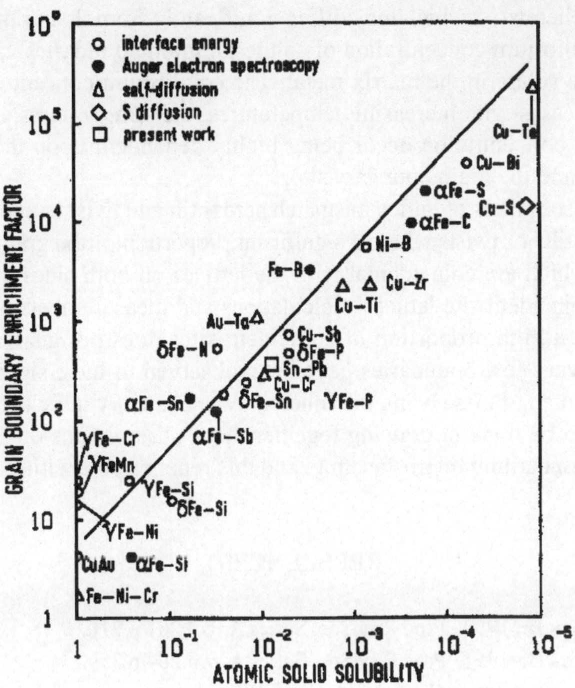

Fig. 1.18 Effect of Solid Solubility on Grain Boundary Enrichment for Different Solutes.

absence of such an energy supply, the rate of migration of the grain boundary will be limited by the diffusion rate of the grain boundary solute. Also the presence of the solute may in fact reduce the grain boundary energy. The effects of solutes is described in Section 3.4.

1.7 SUMMARY

The qualitative description of grain boundaries may not have changed dramatically in the last fifty years, in that a boundary separates two crystals of differing lattice orientation, and the arrangement of atoms in the boundary is disordered over a distance of three or four inter-atomic spacings. The degree of disorder is not uniform, in that the disregistry between the lattices may be minimised by the introduction of either lattice dislocations or secondary grain boundary dislocations.

The very significant advances that have been made over the last 50 years concern developments in electron and other microscopies, which have made the direct observation of atom positions in grain boundaries possible. Examples of high resolution transmission electron micrographs have been illustrated; field ion microscopy is also capable of giving atomic resolution, and, with time-of-flight spectroscopy, is further capable of identifying the individual atoms. Information from high resolution microscopy has shown that boundaries may be non-planar, in that high angle boundaries may decompose into micro-facets, presumable having lower grain boundary energies. Such studies have also shown that the average inter-atomic spacing in the boundary is only slightly greater that that in the crystal. The development of Auger and other spectroscopies has enabled the determination of grain boundary chemistry, which may differ significantly from that of the bulk materials. In general, the equilibrium concentration of solutes at grain boundaries varies inversely as the solubility of the solute in the matrix metal. The equilibrium concentration at the grain boundary also decreases with increasing temperatures, the temperatures at which the most marked changes in concentration occur being highly dependent upon the binding energy between the solute and the grain boundary sites.

Mathematical descriptions of lattice mismatch across tilt and twist boundaries have shown that there are some tilts or twists where a significant proportion of the grain boundary atoms occupy positions which are coincidental with the lattices on both sides of the boundary – these are termed coincident site lattices. Calculations and measurements of grain boundary energies show that a high proportion of coincident sites does not guarantee a low grain boundary energy. Very few boundaries have been observed to have significantly reduced energies, the most noted of these being the coherent twin boundary in f.c.c. metals.

Much remains to be done in drawing together these observations on grain boundaries, particularly those concerning micro-faceting, and this remains an exciting field of study.

REFERENCES

Aust K.T. and Chalmers B., 1950. Proc Roy Soc, Series A, vol.201 p.210.
Aust K.T. and Chalmers B., 1951. Proc Roy Soc, Series A, vol.204 p.359.
Bragg W. L., 1953. Dislocations in Crystals, McGraw-Hill.
Burgers J. M., 1939. Dislocations in Crystals, McGraw-Hill.
Chalmers B., 1937. Proc Roy Soc, Series A, vol.162, p.20
Desch C. H., 1919. Journ Inst Metals, vol.22, p.241.
Dunn C.G, and Lionetti F., 1949. Journ of Metals vol.1, p.125.
English T.H. and Gladman T., 1986. Mat Sci and Tech, vol.2, p.88.
Erb U. and Gleiter H., 1979. Scripta Met, vol.13, p.61.
Ewing J. A. and Rosenhain W., 1900. Phil Trans Roy Soc, Series A vol.193, p.353.
Gladman T., 1997. "The Physical Metallurgy of Microalloyed Steels", Inst of Metals, London.
Goodhew P.J., 1979. Met Sci, March/April, p.108.
Gottstein G. and Schwarzer F., 1992. Materials Science Forum, vols.94-96, p.187.
Grignon, 1776. Memoires de Physique, Paris.
Hargreaves F. and Hills R. J., 1929. Journ Inst Metals, vol.41, p.257.
Harker D. and Parker E.R., 1945. Trans Amer Soc Metals, vol.34, p.156.
Hasson G. C. and Goux C., 1971. Scripta Metall, vol.5, p.889.
Jeffries Z. and Archer E.S., 1924. Science of Metals, McGraw Hill, p.73.
Ke T. S., 1949. Journ app Phys, vol.22, p.274.
Kelvin Lord, 1887. Phil Mag, vol.24, p.503.
Kelvin Lord, 1894. Proc Roy Soc, vol.55, p.1.
Merkle K. L, Wolf D., 1992. Materials Science Forum, vols.94-96, p.475.
Mott N. F., 1948. Proc Phys Soc, vol.60, p.391.
Read W. T. and Shockley W., 1950. Phys Rev, vol.78, p.275
Rosenhain W. and Humphrey J., 1913. Journ Iron Steel Inst, vol.87, p.219
Sauter H, Gleiter H. and Barro G., 1977. Acta Metall, vol.25, p.457.
Schvindlermann L. S. and Straumal B. B., 1985. Acta Metall, vol.33, p.1735.
Seah M. P. and Hondros E. D., 1973. Proc Roy Soc, Series A, vol.335, p.191.

Seeger A. and Schotky G., 1959. Acta Met, vol.7, p.495.
Shuttleworth R., 1952. Imperfections in Nearly Perfect Crystals, Ed. W. Shockley , Wiley, p.343.
Slater J. C., 1939. Introduction to Chemical Physics, McGraw-Hill, p.275.
Smith C. S., 1948. Trans Amer Inst Min Metall Eng. Vol.175, p.15.
Smith C. S., 1952. Metal Interfaces, ASM, Cleveland, Ohio, p.65.
Sorby H. C., 1887. Journ Iron Steel Inst, vol.1, p.255.
Van der Merwe J. H., 1950. Proc Phys Soc, vol.63A, p.616.

CHAPTER 2

Granular Arrays – Topology, Grain Shape, Stereology and Grain Size Measurement

2.1 TOPOLOGY

2.1.1 General Principles

Many of the early concepts of grains and grain shapes were developed from two dimensional models, and were subsequently applied to three dimensional granular arrays, with suitable adaptations for the increased geometrical constraints. For a two dimensional array of polygons, there is a simple relationship between the number of polygons, the number of edges where two polygons meet, and the number of corners where three or more polygons meet. This is given as –

$$P - E + C = 1 \qquad \text{(equ. 2.1)}$$

where P, E, and C are the numbers of polygons (or other flat shapes), edges (lines) and corners (points) respectively. This equation is true irrespective of the number of polygons meeting at a corner, or whether the array includes straight or curved lines. The addition of a point on any edge not only increases the number of points, but also increases the number of edges by unity. The addition of a new curved line between two existing points is the array not only increases the number of edges, but also increases the number of enclosed areas (not necessarily polygons) by unity. This is true even if the curve begins and ends at the same point in the array. These observations ensure compliance with equation 2.1. By way of example, consider the array of shapes shown in Fig. 2.1. The number of points is 57, the number of edges is 44, and the number of polygons or other enclosed shapes is 14, so that, from equ. 2.1 –

$$14 - 57 + 44 = 1$$

It is interesting to compare equation 2.1 with Euler's Law, which describes the relationship between the number of edges, corners, and polygons on the surface of a polyhedron. Euler's Law (C. S. Smith, 1952) is given by –

$$P - E + C = 2 \qquad \text{(equ. 2.2)}$$

This law holds for any polyhedron irrespective of the number of edges meeting at a corner. Take, for example, the simple cube, where three edges meet at each corner. There

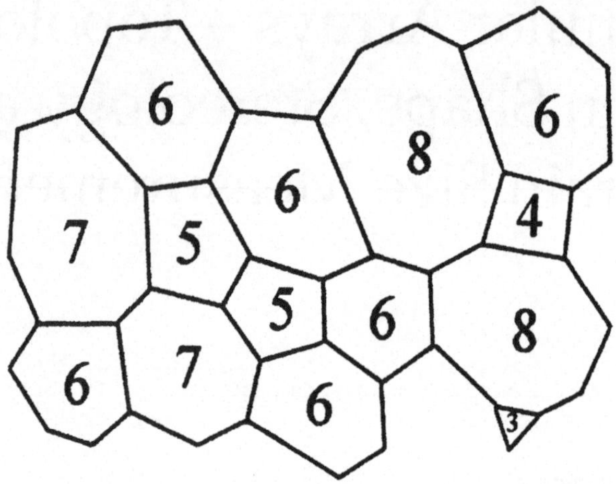

Fig. 2.1 Two Dimensional Polygons. Illustrating the topological equation – P - E + C = 1.

are 6 square faces, 12 edges, and 8 corners, so that 6 - 12 + 8 = 2, in conformity with Euler's Law. For a regular octahedron, where four edges meet at a corner, there are 8 triangular faces, 12 edges, and 6 corners, so that conformity with Euler's Law is again shown, i.e. 8 – 12 + 6 = 2.

Although the surface of a polyhedron cannot strictly be regarded as a two dimensional feature, each facet is two dimensional. The origin of the difference between equation 2.1 and Euler's Law (equation 2.2) is best understood by considering a Schlegel diagram, which was developed to represent the topological features of various polyhedra in a two dimensional drawing. The Schlegel diagram (Aboav D. 1992) is constructed by drawing a frame, having the same number of edges, (and corners) as one of the polygons forming a single face of the polyhedron. Looking into the polyhedron through this particular face, will reveal all of the interconnected faces of the polyhedron, and a portrayal of these will give a topologically correct Schlegel diagram. The frame of the Schlegel diagram corresponds with the boundary of the two dimensional array of polygons and also represents an additional polygonal face of the polyhedron. Thus any two dimensional array of shapes will conform with Euler's Law if the boundary of the array is considered to be an additional polygon. Consider the tetrakaidecahedron shown as a perspective drawing in Fig. 2.2(a). This is also known as a cubo-octahedron, or as Kelvin's fourteen plane faced solid. This shape has eight hexagonal faces and six square faces. Two Schlegel diagrams can be constructed, depending on whether the shape is viewed through one of the square faces, or one of the hexagonal faces. The Schlegel diagram produced by looking through an hexagonal face is shown in Fig. 2.2(b). If Fig 2.2(b) is viewed as a two dimensional array of polygons, there are 6 square faces, but only 7 hexagons. There are 36 edges and 24 corners, so that from equation 2.1, 13 – 36 + 24 = 1. If, however, we regard Fig. 4.2(b) as a Schlegel diagram, then the hexagonal frame itself constitutes an additional face of the polyhedron without the addition of any more edges or corners. Thus, from Euler's Law, 14 – 36 + 24 = 2.

The general topological relationship for an array of polyhedra (Smith 1952), is given as –

$$P - E + C - B = 1 \qquad \text{(equ. 2.3)}$$

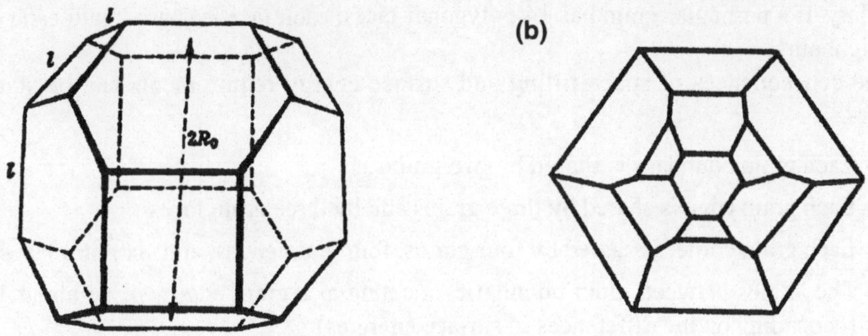

Fig. 2.2 The Tetrakaidecahedron. (a). Perspective view showing the definition of 'radius', the edge length being l. (b). The Schlegel diagram for this shape.

Where P E and C have the same meanings as before, and B is the number of polyhedral grains (or solid bodies) in the array. This is relevant to any array of polyhedra, irrespective of how many faces meet at an edge, or how many edges meet at a corner. It is interesting to note that equation 2.3 can be used for a single polyhedron. In this case, where B = 1, equation 2.3 becomes identical with Euler's Law. Further, if there are no polyhedra (grains, or bodies) then equation 2.3 reverts to the equation for a two dimensional array of polygons.

Whilst these topological rules are universally correct, they do not describe the size or shape of grains, but only the interconnectedness of points, edges, faces, and bodies in space. The absence of scale and angles from the topological relationships means that other considerations are needed when dealing with the shapes and sizes of grains or other arrays of polyhedra.

2.1.2 Surface Energy and Space Filling Effects

In Chapter 1, the nature of grain boundaries was described, and the concept of grain boundary energy (surface tension) was introduced. It was also suggested that many grain boundaries in a single phase material may have very similar grain boundary energies. Relatively few grain boundaries would have significantly lower energies, the most notable being the symmetrical twin boundary so often seen in f.c.c.metals. It was further shown, Section 1.5, that as a consequence of most grain boundaries having very similar energies, certain grain boundary configurations would be inherently unstable. Specifically, a four-rayed edge junction would decompose to form two three-rayed junctions with an accompanying lowering of energy. The lowest energy position corresponded with the angles in the triple points reaching 120°C, i.e. all angles equal. However, this situation is dependent upon equal boundary energies, and variations in grain boundary energy should result in variations in boundary angles as suggested by Shuttleworth (Equation 1.10). Although the configuration shown in Fig. 1.14 is portrayed in two dimensions, with line lengths being minimised, it's relevance to a three dimensional situation can be understood by regarding the lines as intersections between grain boundaries and the plane of the diagram, and the junctions as intersections between grain edges and the plane of the diagram, the boundaries and edges being normal to the plane of the diagram.

An additional constraint is imposed by the space filling requirement. In any solid block of metal, each grain is in contact with it's immediate neighbours. Between each pair of neighbours there is an interface, or grain boundary, and two neighbours share each boundary. If a particular grain has 14 polygonal faces, each face is shared with each of the 14 neighbours.

The consequences of space filling and surface energy requirements can be stated as follows –

a) Each grain boundary is shared by two grains.
b) Each grain edge is shared by three grains and by three grain faces
c) Each grain corner is shared by four grains, four grain edges, and six grain faces.
d) The angles between grain boundaries meeting at a grain edge will be about 120ºC (depending on the differences in surface energies).
e) The angles between grain edges meeting at a corner will be about 109º28'.

It should be noted that these rules for the array are different from those that relate to a single polyhedron where 3 edges meet at a corner, three faces meet at a corner, and two faces meet at an edge.

2.1.3 Implications for Grain Shape

The topological relationships can be modified in the light of the space filling and surface tension requirements to give some idea of grain shape (C.S. Smith 1952). In practice, it is found that grains within an array have different shapes and are of different sizes. The application of the surface energy, space filling, and topological rules can nevertheless be applied to such heterogeneous arrays to give information of the 'average' grain features, (C. S. Smith 1952), as follows.

The basic equation for a three dimensional network of cells has already been cited (equation 2.3) as –

$$C - E + P - B = 1 \qquad \text{(equ. 2.3)}$$

where C is the number of grain corners
E is the number grain edges
P is the number of polygonal faces
and B is the number of polyhedral bodies.

Since each edge is shared by 3 polygonal faces, a polygon having n sides will contribute n/3 edges to the array, so that –

$$E = (\Sigma\, nP_n)/3 \qquad \text{(equ. 2.4)}$$

where P_n is the number of polygon faces having n sides.

Now each edge joins two corners and each corner joins four edges, so that –

$$C = 2E/4 = E/2 \qquad \text{(equ. 2.5)}$$

Substituting equation 2.5 into equation 2.3 gives –

$$P - E/2 = B + 1 \qquad \text{(equ. 2.6)}$$

And further substituting equation 2.4 into equation 2.6 gives –

$$P - \Sigma nP_n/6 = B + 1 \qquad \text{(equ. 2.7)}$$

Now $P = \Sigma P_n$, and the mean value of n, which we will designate **n**, is given by –

$$\mathbf{n} = \Sigma nP_n/\Sigma P_n \qquad \text{(equ. 2.8)}$$

Substituting into equation 2.7 gives –

$$P - \mathbf{n}P/6 = B + 1 \qquad \text{(equ. 2.9)}$$

$$\text{or } (6 - \mathbf{n})\, P/6 = B + 1 \qquad \text{(equ. 2.10)}$$

The average number of polygonal faces per polyhedral body, P/B, is therefore given by –

$$P/B = [6/(6 - \mathbf{n})]\,(1 + 1/B) \qquad \text{(equ. 2.11)}$$

This equation is quite rigorous and there is no correction required for grains, faces, edges, or corners, at the edge of the array. As the number of bodies (grains) increases 1/B approaches zero, and the number of polygonal faces per body (grain) can be closely approximated by –

$$P/B = 6/(6 - \mathbf{n}) \qquad \text{(equ. 2.12)}$$

Dividing equation 2.6 by B gives -

$$P/B - C/B = 1 + 1/B \qquad \text{(equ. 2.13)}$$

Again, as B increases, 1/B approaches zero, and equation 2.13 becomes –

$$C/B = P/B - 1 \qquad \text{(equ. 2.14)}$$

Substitution for P/B from equ. 2.12 yields –

$$C/B = 6/(6 - \mathbf{n}) - 1 = \mathbf{n}/(6 - \mathbf{n}) \qquad \text{(equ. 2.15)}$$

Thus the average number of polygonal faces per body (grain), P/B, and the average number of corners per body, C/B, in an array of polyhedral are clearly expressed in terms of the average number of corners per face by equations 2.12 and 2.15 respectively, and these two numbers will differ by unity, as expressed in equation 2.14.

It is clear from these equations that the average number of corners on the polygonal faces cannot exceed 6, and it is interesting to examine the implications of equations 2.12 and 2.15 on the shape of metal grains. First, it must be recognised that the values of C/B and P/B relate to the array as a whole. If an individual body (grain) is isolated from the array, then the number of polygonal faces must be doubled as each face is shared by two adjacent grains, and the number of corners must be increased by a factor of four, as each corner is shared by four grains. Smith (1952) drew attention to an array of polyhedra having C/B = 6, P/B = 7, and with an average of $5^1/_7$ corners per polygonal face, i.e. an array of grains with an average of 14 polygonal faces, and an average of 24 corners. The tetrakaidecahedron with its 8 hexagonal faces and 6 square faces complies with these requirements exactly, i.e. there are 14 faces, 24 corners, and an average of $5^1/_7$ corners for each polygonal face. Further, uniform tetrakaidecahedra can be stacked to fill space, provided that they are stacked on a body centred cubic space lattice, Fig. 2.3, (C. S. Barrett and J. B. Massalski, 1980). In this array, three grains meet at an edge and four edges meet at a corner, as required by surface tension effects, but the local geometry at the grain edges and corners does not conform with the angular requirements of 120° between faces at a grain edge and 109°28' between grain edges at a corner. However, it has been shown (Lord Kelvin, 1887 and 1894) that the local angular equilibrium can be maintained by introducing doubly curved boundaries as shown in Fig. 2.4.

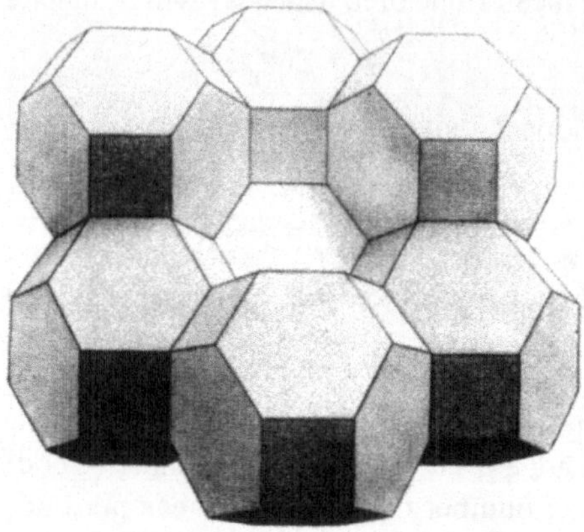

Fig. 2.3 Stacked Tetrakaidecahedra. The shapes are stacked on a body centred cubic space lattice.

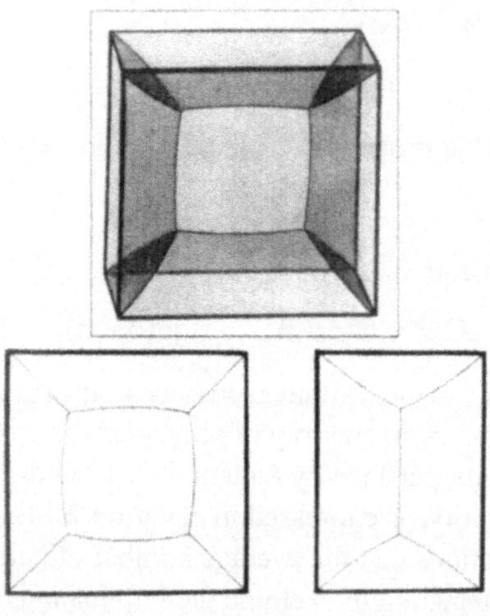

Fig. 2.4 Kelvin's Doubly Curved Surfaces. Illustrated by soap films on a wire frame.

The compliance of the array of uniform tetrakaidecahedra with space filling, and with equations 2.12, 2.14, and 2.15, however does not prove that grains have this shape. There is no requirement in the topological rules or in the space filling requirements for grains to have either the same shape or the same size. The values given in Table 2.1 for the number of faces and corners for the grains in an array are average values, and as such may represent many different arrays of grains of differing shapes. Neither do the topological and surface energy requirements dictate that C/B should be 6 (and therefore that P/B should be 7) as other options are shown in Table 2.1. With C/B = 5, a regular shape dictated by P = 12 would be the pentagonal dodecahedron, Fig. 2.5, with it's angles between edges of 108°. Unfortunately, this figure lacks the dimensional symmetry to fill space, but this does not rule out a mixture of grains with an average of 12 faces and an average of 5 corners per polygonal face.

Nevertheless, despite the above reservations, characteristics of the tetrakaidecahedron have made it a useful shape for mathematical modelling of grain structures.

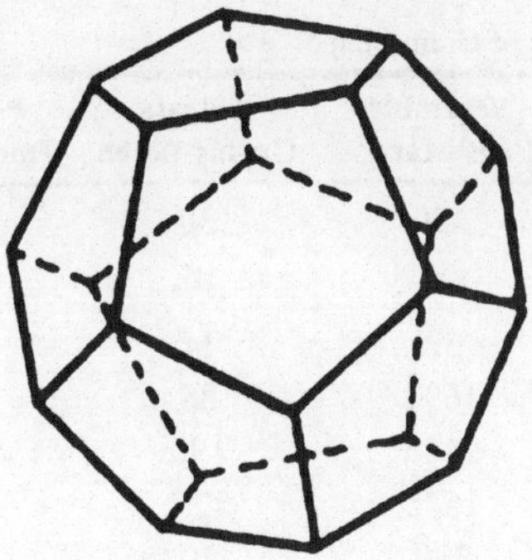

Fig. 2.5 The Pentagonal Dodecahedron. This shape has 20 corners that could represent the carbon atoms in Buckmaster Fullerene.

2.1.4 Experimental Assessment of Grain Shape

Granular, or cell, structures are encountered in a variety of materials, and have stimulated considerable interest ever since Grignon sketched his iron grains in 1776. Measurements of the cell/grain shapes have mainly been conducted by counting the number of edged on each grain face. Other parameters can of course be derived using Equations 2.4 and 2.15. The most extensive work has been carried on soap-froths (Matzke, 1946, Matzke and Nestler 1946), but other work has been carried out on vegetable cells (Matzke 1946) and on β-brass (Desch 1919). The appearances of the granular structure of alumina and a soap froth are compared in Fig. 2.6. Despite the fact that the metallic structure is obtained from a thin film of alumina, whilst the soap froth is several cells deep, the similarity is clearly seen.

The frequency distributions of the number of corners per polygonal face for the various materials, and also for two standard polyhedra, are shown in Table 2.2. An important feature of the experimentally determined frequency distributions is that the most commonly occurring grain face has five sides. Various shape characterising parameters derived from the frequency distributions are also shown in Table 2.2. The average number of corners per grain face, **n**, was calculated from equation 2.8, for each of the materials. For the soap froth, the vegetable cells, and the β-brass grains, the average number of corners per face was very close to that for the tetrakaidecahedron, much closer than to that of the pentagonal decahedron. The experimentally observed average number of faces per grain for the cells/grains, obtained from $P/B = \Sigma P_n/B)$, have values ranging from 6.85 to 7.25

Table 2.1 Topological Relationship for Polyhedra in a Space Filling Array.

Corners per Body C/B	Faces per Body C/B	Mean No. of Edges per Face, n	Faces/Single Grain = 2P/B	Corners/Single Grain = 4C/B	Regular Polyhedron with These Characteristics
1	2	3	4	4	Tetrahedron
2	3	4	6	8	Cube
5	6	5	12	20	Pentagonal Dodecahedron
6	7	$5^{1/7}$	14	24	Tetrakaidecahedron
7	8	$5^{1/4}$	16	28	—
8	9	$5^{1/3}$	18	32	—
Infinite		6			

Note – Each face shared by two bodies in the array, and each corner shared by four bodies.

Table 2.2 Experimental Observations of Grain Shape

Material	**Soap Froth Matzke**	**Vegetable Cells Matzke**	**β-Brass Grains Desch**	**Pentagonal Dodecahedron**	**Tetrakaidecahedron**
Cells Counted	600	450	30	1	1
n	$2P_n$	$2P_n$	$2P_n$	$2P_n$	$2P_n$
3	0	32	11	0	0
4	866	1694	88	0	6
5	5503	2465	190	12	0
6	1817	1575	123	0	8
7	35	390	20	0	0
8	0	50	3	0	0
9	0	5	0	0	0
$\Sigma 2P_n$	8221	6211	435	12	14
$\mathbf{n} = \Sigma n2P_n/\Sigma 2P_n$	5.111	5.123	5.142	5.000	5.143
Corners/Body C/B	5.851	5.893	6.213	5.000	6.000
Faces/Body P/B	6.851	6.901	7.250	6.000	7.000
Corners/Single Grain	23.40	23.56	24.84	20.00	24.00
Faces/Single Grain	13.70	13.80	14.50	12.00	14.00
P/B-C/B	1.000	1.008	1.037	1.000	1.000
Expected Value	1.002	1.002	1.033	—	—

which compare well with the value 7.00 for the tetrakaidecahedron and certainly exceed the value of 6.0 for the pentagonal dodecahedron. The average number of faces associated with a single grain would therefore appear to be very close to 14 (values of 13.70 to 14.50). Similar remarks can be made about the average number of corners per cell/grain, which again resemble that of the tetrakaidecahedron each cell or grain having a average of between 23.4 and 24.84 corners.

It is clear that the combination of topological, surface energy and space filling requirements result in an average grain shape which conforms closely with the parameters of the tetrakaidecahedron in having about 14 faces, and about 24 corners, but real grains differ from the tetrakaidecahedron in two very important respects. In a real granular array, the five-sided grain face is the most common, Fig. 2.6, whereas there is no five sided face in the tetrakaidecahedron. Uniform tetrakaidecahedra can only fill space by occupying a body centred cubic space lattice, whereas a real granular array fills space by having different grain shapes with shared interfaces, thus giving corner and edge configurations that show better conformity with surface tension requirements than the bimodal local configurations provided by the tetrakaidecahedron.

In summary, the experimental evidence points to an 'average grain' which has 14 faces each shared by two grains giving 7 faces per grain. Also the average grain will have 24 corners, each shared between four grains, giving 6 corners per grain. Although, these average values are very close to those of Kelvin's 14 plane faced solid, the most commonly observed grain face has five sides, and the granular array does not require the regularity provided by a single space filling shape. This variation in grain shape is an important feature of granular arrays, for it is the irregularity that allows grain growth to occur when the temperature is high enough for this thermally activated process to occur.

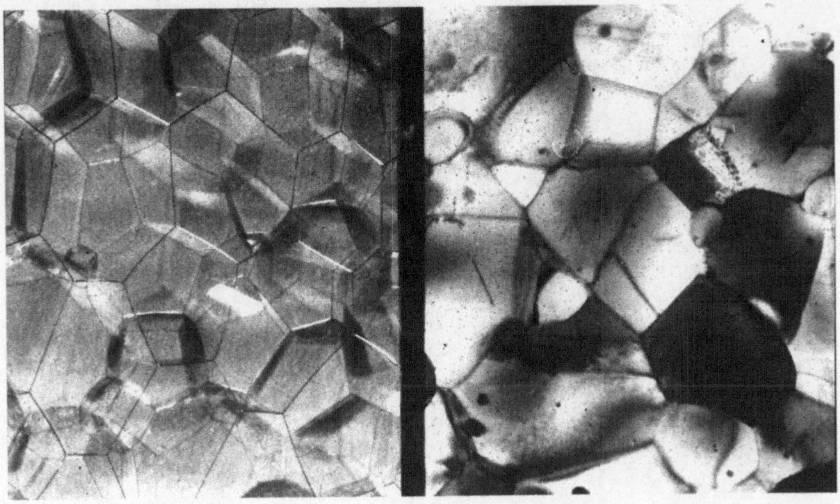

Fig. 2.6 Comparison of a Soap Froth with the Granular Structure in a Fine Grained Alumina. Only the magnification is changed.

2.2 GRAIN SIZE

The huge importance of grain size on the properties of metals has already been addressed in the introduction. The importance of grain size has led to considerable interest in controlling the grain size of metal, whether this is to encourage the development of extremely large grains as in magnetic materials, or to encourage the development of extremely small grained metals, with benefits to both the strength and toughness, particularly in ferrous (or other b.c.c.) alloys. Not unnaturally, the great importance of the control of grain size, and, consequently of the control of material properties, has led to a need for methods of grain size characterisation and measurement, for both research and quality control purposes. There are two main problems associated with grain size measurements, the first being associated with the selection of a suitable measurement parameter, and the second arising from the fact that metals are opaque, and the three dimensional grain structures cannot often be observed directly. The difficulties in selecting a measurement parameter can be appreciated from the photographs of a soap froth shown in Fig. 2.7. Potential 3-D parameters would include grain volume, grain edge length, grain face area, or grain calliper diameter. The main difficulty is to identify the specific cell shape. This would certainly be made easier with a stereographic image as was used by Williams (1952), but the production of a satisfactory stereographic image in that case was assisted by the presence of a dense second phase on the grain boundaries, and this is not generally present in commercial materials. If specific grains could be seen, the measurement of grain volume could easily be derived by counting the number of grains per unit volume, given that the overall dimensions of the sampled volume were known.

However, direct measurements of this kind are irrelevant in opaque metals. Indirect methods such as X-ray methods have been used, and an excellent account of the use of X-rays for grain size determination has been given by Andrews (1973). Such methods still require some degree of specimen preparation, the interpretation can be lengthy, and X-ray

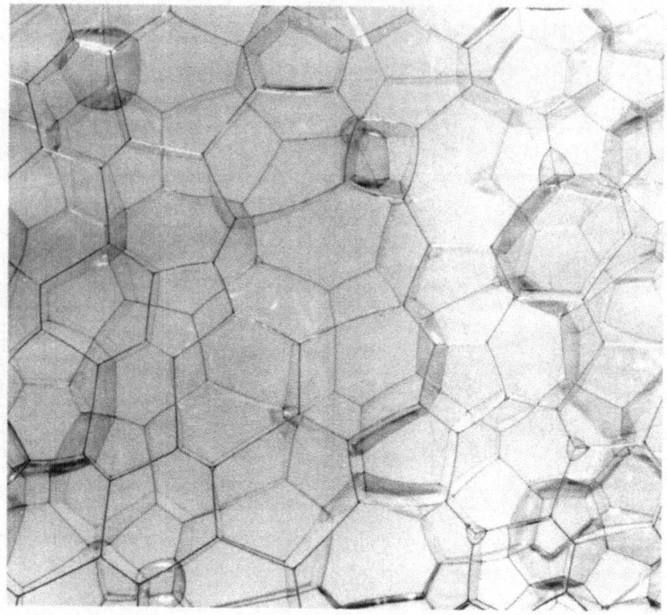

Fig. 2.7 A Three Dimensional Soap Froth. How would you measure the size of the bubbles?

methods necessarily involve considerable safety precautions. The most common methods for grain size assessment involve metallographic measurements made on planar sections. Before describing the methods used, it is useful to give some consideration to stereological principles, which relate observations made on a two dimensional section to the characteristics of the three dimensional structure.

2.2.1 Stereological Principles

Measures of grain size that are appropriate to a planar section involve either areal or linear measures. The general stereological relationships that exist between the different measures made on a plane section and the corresponding quantities in three dimensions are summarised in Table 2.3, (Underwood, 1970).

The first column in Table 2.3 relates to the determination of the volume fraction of a particular phase in a multiphase structure. Although this has no direct bearing on grain size, it is included here because it is relevant to the determination of grain size in multiphase structures. The volume fraction of α-phase, f_α, in a structure cannot be determined directly

Table 2.3 Stereological Relationships.

Dimensions	—	mm^{-1}	mm^{-2}	mm^{-3}
Points	P_P	N_L	N_A	N_V
Lengths	$L_L=P_P$	$L_A=\pi N_L/2$	$L_V=2N_A$	—
Areas	$A_A=P_P$	$A_V=2N_L$	—	—
Volumes	$V_V=P_P$	—	—	—

Note: (a) Surface area per unit volume = $A_V = 2N_L = 4L_A/\pi$;

by metallographic methods, but is numerically equal to the fractional area occupied by the α–phase in a plane section, or the fractional length occupied by the α-phase on a line in such a plane, (Delesse 1848, Rosiwal 1903). In the case of the area fraction –

$$f_\alpha = A_\alpha / A \quad \text{(equ. 2.16)}$$

Where A_α is the total area of α-phase in a total area A. Limited use was made of this relationship due to the effort required to take micrographs of the structure and then cut out and weigh the areas denoting the two phases. The use of lineal analysis to determine the fractional line length occupied by α-phase i.e.

$$f_\alpha = L_\alpha / L \quad \text{(equ. 2.17)}$$

was greatly facilitated by instruments with several micrometers, each one assigned to a given phase, but all contributing to the linear motion of the sample under a reference crosswire. The length of line in each phase could be noted directly. The use of point counting has probably been the most popular method, involving the projection of n random points into a microstructure, and determining the number lying in the α-phase, n_α, when –

$$f_\alpha = n_\alpha / n \quad \text{(equ. 2.18)}$$

Lineal analysis and point counting were used almost exclusively in metallurgical work for volume fraction assessment over a period of about 60 years, until the advent of reliable image analysis equipment in the 1970's. The basis of volume fraction determination in this case is really a matter of computer programming. Image analysis equipment however requires a substantial capital investment, and manual methods are still in use, particularly within smaller organisations.

The statistical errors associated with point counting (Gladman and Woodhead, 1960) and lineal analysis (Gladman 1963), have been quantified. In the case of point counting, provided that the points are randomly distributed with respect to the microstructure (or that the microstructure is randomly distributed with respect to a regular array of points), then the standard deviation, σ_f, of the volume fraction of α-phase, f_α, is given by –

$$\sigma_f = [f_\alpha (1 - f_\alpha)/n]^{1/2} \quad \text{(equ 2.19)}$$

This equation will relate to any multiphase system irrespective of the number of phases. The error is symmetrical with respect to f_α, increasing as f_v increases from $f_\alpha = 0$ to $f_\alpha = 0.5$, and then decreasing from $f_\alpha = 0.5$ to f_α - 1.0.

The standard deviation of an estimated volume fraction obtained by lineal analysis, was derived by Hilliard and Cahn (1961) as –

$$\sigma_f = [f_\alpha (1 - f_\alpha)/n^{1/2}]\,[(\sigma_i/i)_\alpha{}^2 + (\sigma_i/i)_\beta{}^2]^{1/2} \quad \text{(equ 2.20)}$$

where σ_f is the standard deviation of the mean intercept length
i is the mean intercept length for the phase denoted (α or β)
and n is the number of α-phase particled intersected.

Subsequent work by Gladman (1963) showed that the distribution of intercept lengths was such that the standard deviation of the intercept length σ_i, was equal to the mean intercept length, i, so that equation 2.20 may be simplified to –

$$\sigma_f = f_\alpha (1 - f_\alpha)(2/n_\alpha)^{1/2} \quad \text{(equ. 2.21)}$$

It should be noted that the standard deviation of the volume fraction is again a symmetrical function of the volume fraction over the full range from 0 to 1, but the standard deviation of the volume fraction depends on the number of α-phase regions intersected rather than on a specific line length.

The statistical error associated with areal analysis was investigated by Hilliard and Cahn (1961), who assumed, for the purpose of the model, that the α-phase occurred as discrete second phase particles. On this basis, they showed that –

$$\sigma_f = f_\alpha [1 + (\sigma_a/a)^2]^{1/2} / n_\alpha^{1/2} \qquad \text{(equ. 2.22)}$$

where σ_a is the standard deviation of the individual α particle area
a is the mean area of the α particles measured
and n_α is the number of α particles measured.

There has been little examination of Equation 2.22 but it is probable that σ_a is a reasonable approximation of a, and therefore Equation 2.22 could be approximated by –

$$\sigma_f = f_\alpha (2/n_\alpha)^{1/2} \qquad \text{(equ. 2.23)}$$

However, Equations 2.22 and 2.23 lack the symmetry of Equations 2.19 and 2.21, and imply that the standard deviation of f_α is different from the standard deviation of $(1 - f_\alpha)$. This is clearly untrue.

The errors associated with automatic image analysis would thus depend on the computation of the area fraction itself. It would be incorrect to equate the number of pixels with points in a point counting system, there will be very many pixels lying within any one particle of α-phase, but a satisfactory image can only be produced by having many pixels within an area of α-phase. The consequence is that the pixels cannot be regarded as independent and the errors cannot therefore be obtained from equation 2.19. Linear intercepts from a row of pixels would not suffer the same problem because linear intercepts are based on line lengths, which is estimated from the number of pixels lying in the same particle. It is likely that the error is described by an equation similar to equation 2.21 or 2.23. If, as was assumed by Hilliard and Cahn, the volume fraction of α-phase is very small, then f_α is a reasonable approximation of $f_\alpha(1 - f_\alpha)$. The ease of operation of automatic equipment enables larger statistical samples to be taken, and the standard deviation to be evaluated from multiple estimates i.e. –

$$\sigma_f = \Sigma(f_i - f_\alpha)^2 / n - 1 \qquad \text{(equ. 2.24)}$$

where f_i is the 'i' th determination
f_α is the mean value of f_i
and n is the number of determinations of f_i.

The remaining stereological expressions shown in Table 2.3, are more related to the description of grain size or other scaled features. The three parameters associated with volume measures, i.e. area per unit volume, A_v, length per unit volume, L_v, and number per unit volume, N_v, cannot be determined directly, but have to be derived from the other measures as indicated in the table. The number per unit length is a simple count of the number of times a feature is intersected by a linear traverse (or a line of known length drawn on a micrograph). Although the lengths per unit area can be derived from –

$$L_A = \pi N_L/2 \qquad \text{(equ. 2.25)}$$

there are limitations to the use of this relationship. If, for example, the number of grain boundaries per unit length of an intersecting line is counted then equation 2.25 is relevant, provided that the grain structure is isotropic. If the grains are elongated, then equation 2.25 would only hold if the number of features were counted on a curved line oriented randomly in the plane of the microstructure, e.g. a circle, or on a series of straight lines representing all possible orientations. In a non-oriented structure (or using an appropriate counting system) the length of grain boundaries per unit area of the prepared section can be calculated from a count of grain boundaries intersected per unit length of a linear intercept, (the reciprocal of the mean intercept length), using Equation 2.25.

The calculation of surface area per unit volume from –

$$A_V = 2N_L \qquad \text{(equ. 2.26)}$$

– also requires an isotropic structure, or needs to be sampled with a line representing all directions in 3D, and not just in the plane of polish. The use of Equation 2.26 is therefore only recommended for equiaxed structures. Provided that the grains are equiaxed, then the surface area per unit volume can be obtained simply by doubling the number of surfaces intersected per unit length of the intersecting line.

The length of certain features per unit volume is found useful in the measurement of dislocations, where the total length of dislocation per unit volume can be derived from a count of the number of dislocations exiting unit area of free surface of a thin foil specimen, using electron microscopy, as shown in the following equation –

$$L_V = 2N_A \qquad \text{(equ. 2.27)}$$

Provided that the dislocations are in contrast, it is relatively easy to count all ends of the dislocations. It should be remembered that dislocation images may exit by both the top and bottom surfaces of the foil, and this must be taken into account when evaluating the area counted. Another possible application of this relationship is in assessing the total length of grain boundary edges per unit volume. This can be achieved quite simply by counting the number of grain boundary junctions (triple points) per unit area of a metallographic plane section. The total length of grain boundary edge per unit volume is simply obtained by doubling the number of triple points per unit area.

As shown in Table 2.3, the number of features per unit volume can be obtained from measures of the number per unit area and number per unit length, according to the equation –

$$N_V = 2N_A \times N_L \qquad \text{(equ. 2.28)}$$

If the number of grains occurring in a unit area of a planar section is determined, together with the number of grains intersected by an intersecting line of unit length, the number of grains per unit volume is easily obtained as twice the product of these two measures. This can obviously be used as a measure of grain size, but may be expanded as follows. The volume fraction is given by –

$$f_V = N_V v \qquad \text{(equ. 2.29)}$$

where v is the mean grain volume. In the case of a single phase material, $f_V = 1$, and therefore the mean grain volume is given by –

$$v = N_V^{-1} \qquad \text{(equ. 2.30)}$$

Thus the mean grain volume can be obtained from relatively simple measures of number per unit area and number per unit length made on a simple planar section.

In the case of two phase structures, the same principles are involved but have to be

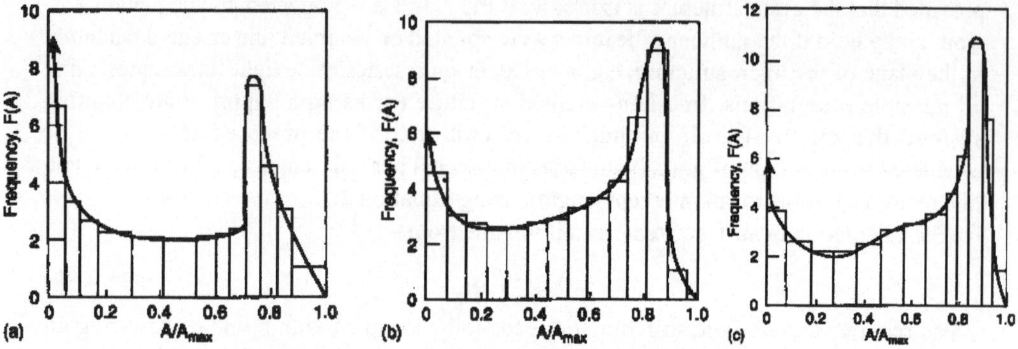

Fig. 2.8 Distribution of Areas of Intersection. (a). A Cube. (b). Tetrakaidecahedron. (c). Pentagonal dodecahedron.

modified to take account of the second phase. For example, if an intercept of length L is superimposed on a two phase microstructure, and the number of α-phase grains N_α and the number of β-grains N_β are counted, the number of α grains per unit length of intercept in the α-phase, $N_{L\alpha}$, is given by –

$$N_{L\alpha} = N_\alpha / L_\alpha \qquad \text{(equ. 2.31)}$$

From Equation 2.17, $f_\alpha = L_\alpha/L$, and substitution into Equation 2.31 gives –

$$N_{L\alpha} = N_\alpha / f_\alpha L \qquad \text{(equ. 2.32)}$$

The mean linear intercept in the α-grains, i_α, is then the reciprocal of $N_{L\alpha}$, and can be expressed as –

$$i_\alpha = f_\alpha L / N_\alpha \qquad \text{(equ. 2.33)}$$

The same treatment can be applied to the β-phase.

Similarly, if an intersecting area, A, is superimposed on a two phase microstructure, and the number of α-grains is counted, N_α, then the number of grains per unit area of α-phase $N_{A\alpha}$ is given by –

$$N_{A\alpha} = N_\alpha / f_\alpha A \qquad \text{(equ. 2.34)}$$

and the reciprocal of this gives the mean areal intercept, a_α, of the α-grains, i.e.

$$a_\alpha = 1/N_{A\alpha} = A f_\alpha / N_\alpha \qquad \text{(equ. 2.35)}$$

Thus the mean linear intercept, i_α, and the mean areal intercept of the α-grains, a_α, has been obtained simply by correcting for the volume fraction. However, the number of α-grains per unit volume is defined by equation 2.28 and uses the unmodified N_L and N_A, i.e. the number of α-grains per unit length of the whole intercept size, and the number of α-grains per unit area of the whole intersecting area, thus giving the number of α-grains per unit volume of specimen. In order to determine the mean volume of the α-grains, v_α, the number of α–grains per unit volume of α-phase $N_{v\alpha}$ is required, and v_α is given from Equation 2.29 as –

$$v_\alpha = N_{v\alpha}^{-1} = N_v^{-1} . f_\alpha \qquad \text{(equ. 2.36)}$$

(note – $N_{v\alpha}$ cannot be obtained simply by substituting $N_{A\alpha}$ and $N_{L\alpha}$ into Equation 2.28).

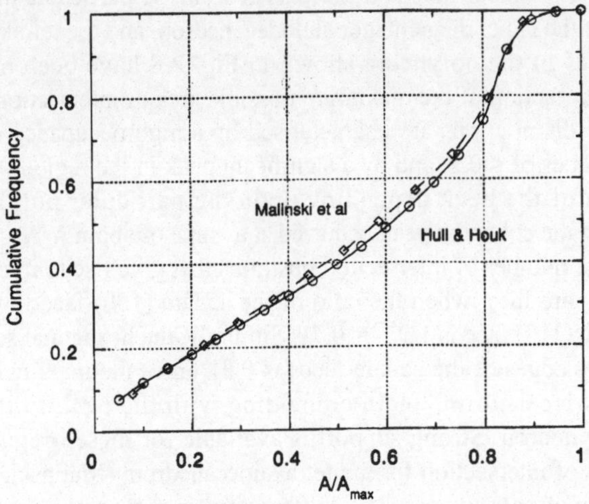

Fig. 2.9 Comparison of Areal Intersections. (Hull and Houk, 1953; Malinsky et al, 1992).

In summary, the stereological principles can be used to relate the mean linear intercept of α-grains, mean areal intercept of α-grains, and the mean volume of α-grains in single and polyphase structures, with various volume fractions of phases. However, in order to try to relate the mean values to other grain dimensions, some attention must be given to the distributions of these mean values.

2.2.2 Stereological Aspects of Various Polyhedra

Studies of the distributions of both the areal intersections and linear intercepts have been carried out for various polyhedra. The aim of this work was to relate the mean areal intersection or the mean linear intercept to the 3-D "grain size". Whilst the size of a sphere can be directly related to the diameter, surface area, or volume, there are several measures that can be used in the case of any polyhedron. For example, the size may be expressed as the distance between diagonally opposite corners, or diagonally opposite faces, both of which may have more than one value. For example, in the tetrakaidecahedron the distance between the opposite square faces is about 29% greater than the distance between opposite hexagonal faces. Other measures that avoid the use of a specific geometry include the volume, or the diameter of a sphere of equivalent volume. In the following consideration of areal and linear intercepts, the relationships between the mean intercept and measures of spatial size will be illustrated.

(A) Mean Areal Intersection

The distribution of areas of intersection between a particular body and the sampling plane (Hull and Houk, 1952), are shown in Fig. 2.8, for various polyhedra. Hull and Houk used wire frame models of the various polyhedra that could be tilted, rotated and elevated. The model was stood in a liquid bath so that the liquid surface represented the plane of intersection. Measurements on the intersection included the number of edges and the area of the intersection. These measures were made at different heights, tilts, and rotations with respect to the intersecting planes, selected to give a representative coverage of all possible intersections between the plane surface and the polyhedron. Of particular interest in their work were the results for the cube, the pentagonal dodecahedron, and the tetrakaidecahedron. The distribution of areas in the polyhedra shown in Fig. 2.8 have been normalised by expressing each area as a fraction of the maximum possible area of intersection for the given polyhedron. All of these distributions are characterised by a high occupancy of the smaller size classes at A/A_{max} values of < 0.2, and by a significant peak in the region $0.7 < A/A_{max} < 0.9$, the exact position of the peak depending upon the particular polyhedron being considered. In the case of the cube, the peak occurred at a value of about $A/A_{max} = 0.7$, which can be related to the large number of intersection positions that give rise to intersection areas just above that of the square face, when the ratio of the square (100) face to the maximum area of intersection of the (110) face is $1/2^{1/2} = 0.71$. Similarly, the hexagonal section through the centre of the opposite edges of the square faces is 0.81 times the maximum (octagonal) section for the tetrakaidecahedron, again coinciding with the peak in the frequency distribution for this polyhedron. Strong support is available for these irregular frequency distributions of the areas of intersection for the tetrakaidecahedron, (Malinski, Cwajian, and Chapronski, 1992). Although carried out in very different ways, and using very different class intervals, cumulative frequency curves were calculated for both sets of results (Gladman, 1997), and are shown in Fig. 2.9. The inflections in the cumulative frequency curves for the tetrakaidecahedron are reproduced in both sets of results and show very close agreement. The distributions shown in Fig. 2.8 can therefore be used with some confidence, and provide a basis for calculating 3-D parameters from the mean areas of intersections.

The relationship between the average intersected area and the maximum area of intersection for the various shapes gives a means of estimating other parameters, as these are uniquely related for a given shape.

These factors are shown in Table 2.4. For a given mean area of intersection the maximum area of intersection increases from the sphere to the tetrakaidecahedron to the cube, as do the grain volume and equivalent grain diameter. Whilst it is usually true that these measures increase as the number of faces of a polyhedron decrease, there are exceptions that depend upon the precise geometry and the symmetry. The interesting feature of the conversion factors in Table 2.4, is that the magnitude of the conversion factors from mean area to equivalent diameter is not very shape sensitive, only varying from 1.424 for the tetrakaidecahedron with its 14 faces to 1.382 for the sphere. Given that the topological evidence suggests grains that on average have 14 faces, it would appear reasonable to use the conversion factor of 1.424 to convert a mean area of intersection (obtained from the number of grains intersected per unit area of the section), to an equivalent diameter, i.e. the diameter of a sphere having the same volume as the tetrakaidecahedral grains intersected.

Table 2.4 Conversion Factors for Average Areal Intercept, A_a, for Different Shapes, to the Spatial Equivalent Diameter, d_{eq}, of a Sphere.

Shape	A_a/A_{max}	A_{max}/A_a	Volume	d_{eq}
Cube	0.467	2.14	$1.861.A_a^{3/2}$	$1.526.A_a^{1/2}$
Tetrakaidecahedron	0.546	1.83	$1.512.A_a^{3/2}$	$1.424.A_a^{1/2}$
Pentagonal Dodecahedron	0.592	1.69	—	—
Sphere	0.667	1.50	$1.382.A_a^{3/2}$	$1.382.A_a^{1/2}$

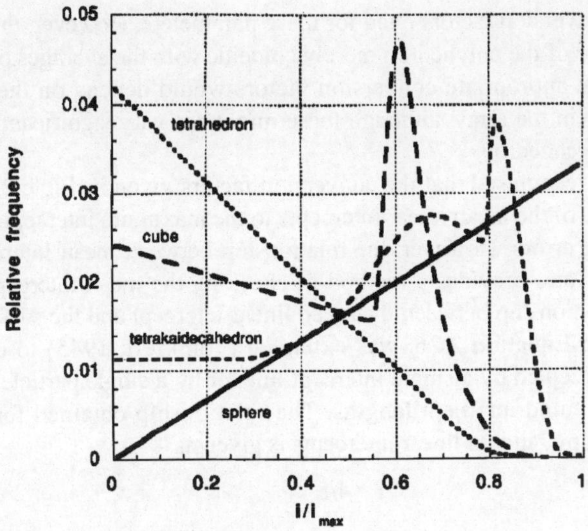

Fig. 2.10 Linear Intercept Distributions for Various Polyhedra. The distribution of intercepts for a sphere are also shown

(B) MEAN LINEAR INTERCEPT

The distribution of linear intercepts from various polyhedra (Gladman, 1997), are shown in Fig. 2.10. The linear intercept distribution obtained for the sphere is also shown for comparison. The distributions for the cube and the tetrakaidecahedron show the same irregularities as were observed for the areal intersections shown in Fig. 2.8. The peaks again show some dependency on the principal dimensions of the particular polyhedron concerned. For example, in the case of the cube, the intercept length normal to the cube face is 0.55 of the maximum intercept lengths, which coincides with the onset of the large peak in the intercept frequencies. The stereological conversion factors required to convert the mean linear intercept, i, into maximum intercept length, grain volume, and equivalent diameter (diameter of a spherical grain of the same volume) are shown in Table 2.5, each of these being derived from the values of i/i_{max} determined from the frequency distributions shown in Fig. 2.10. The equations describing the maximum intercept and volume of the various polyhedra in terms of edge length are also included in Table 2.5. The conversion factors based on intercept lengths are far more shape sensitive than those based on areal intersections. The selection of a particular figure such as the tetrakaidecahedron, with it's 14 faces and average of $5^1/_7$ edges per face would allow the maximum intercept length with a grain, the grain volume, and the equivalent diameter, to be estimated from the mean linear

Table 2.5 Conversion Factors for Mean Linear Intercept, i, for Different Shapes, to the Spatial Equivalent Diameter, d_{eq}, of a Sphere.

Shape	i_{max}	Volume	i/i_{max}	i_{max}/i	Volume/i^3	d_{eq}/i
Tetrahedron	e	$2^{1/2}e^3/12$	0.265	3.77	6.315	2.29
Cube	$3^{1/2}e$	e^3	0.387	2.58	3.305	1.85
Tetrakaidecahedron	$10^{1/2}e$	$128^{1/2}e^3$	0.546	1.83	2.193	1.61
Sphere	d	$\pi d^3/6$	0.667	1.50	1.767	1.50

e= edge length; d = diameter

intercept, using the conversion factors cited for these parameters. However, the number of faces and edges per face of the polyhedron merely coincide with the averages of real arrays of metal grains and the appropriate conversion factors would depend on the mixture of grain shapes that occur in the array, although these might not vary significantly from that shown for the tetrakaidecahedron.

Finally it should be recognised that the conversion factors given in Table 2.5 are based on estimates of the ratio of the mean linear intercepts to the maximum intercept length, and, are subject to sampling errors. However, the relationship between mean intercept lengths and surface to volume ratio provides a method of checking the mean intercepts obtained from Fig. 2.10. The relationship between the mean linear intercept and the surface area per unit volume, shown in Equation 2.26, was extended (Tomkieff, 1945) to cover many separated particles intercepted by a single intercept line, or by a single particle intercepted by many randomly oriented intercept lengths. The relationship obtained for the single shape intercepted by many random linear intercepts is given as –

$$i = 4/S_v \qquad \text{(equ. 2.37)}$$

This relationship is true irrespective of the shape and whether it is a concave or convex particle or whether the surfaces are flat or curved. The proportionality constant is exactly twice that given in Equation 2.26 because the separation of grains or shapes doubles the surface area. In an array of space filling shapes, each grain is shared by the two contiguous grains. Thus, Equation 2.37 gives an unambiguous definition of the mean linear intercept of various shapes from a knowledge of their respective surface areas and volumes.

An assortment of regular polyhedra and the equations for surface areas and volumes are given in Fig. 2.11 (Rade and Westargren, 1992) and data for the tetrakaidecahedron were taken from the work of Rutherford Arborn and Bain, (1937). The circumradii for these shapes are also shown. Using this data, augmented by similar information for the tetrahedron and the tetrakaidecahedron, the surface area to volume ratios were calculated and converted to the mean linear intercepts using equation 2.37. The results are shown in Table 2.6. Also shown are ratios of i/i_{max} obtained by the inter-facial area method, and the corresponding values obtained for certain of the shapes from the frequency distributions of Fig. 2.10.

The surface area to volume ratio generally increases for a given edge length as the complexity of the polyhedron increases, e.g. number of corners. Thus the mean linear intercept increases with the complexity and although the maximum intercept length also increases with this same complexity, there is still a shape sensitivity to the value i/i_{max}. It should be noted however, that the sensitivity to shape is only weak once the polyhedron has

Table2.6 Comparison of Conversion Factors from Eqn.2.37 (from Area /Volume Ratios) with Those from the Frequency Distributions of Fig.2.10.

Shape	A/V	i Eqn2.26	R	i_{max}	i/i_{max} Eqn2.37	i/i_{max} Fig2.10
Tetrahedron	14.7/e	0.27e	0.61e	1.00e	0.27	0.27
Octahedron	7.3/e	0.54e	0.71e	1.41e	0.38	—
Cube	6.0/e	0.67e	0.87e	1.73e	0.38	0.39
Icosahedron	4.0/e	1.01e	0.95e	1.90e	0.53	—
Dodecahedron	2.7/e	1.49e	1.40e	2.80e	0.53	—
Tetrakaidecahedron	2.4/e	1.69e	1.58e	3.16e	0.53	0.53

e = edge length; R = circumradius

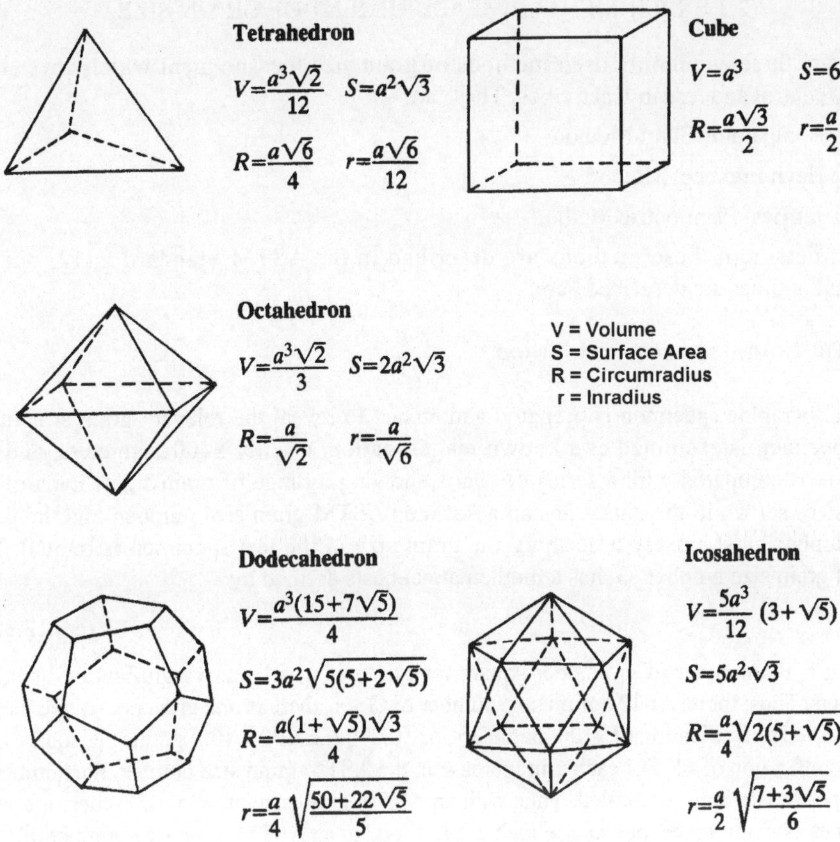

Fig. 2.11 Geometrical Aspects of Regular Polyhedra.

a substantial number of faces/corners. The ratio i/i_{max} is 0.53 for the icosahedron, the pentagonal dodecahedron and the tetrakaidecahedron, so that for the majority of real grains the maximum intercept or "diameter" can be obtained from the mean intercept using the reciprocal of this conversion factor.

The values of i/i_{max} obtained from the frequency distributions of Fig. 2.10 show very good agreement with the absolute values obtained from Equation 2.37. This gives strong support to the computer methods used for sampling the intercept lengths, and for the intricacy of the frequency distributions obtained for the polyhedra so far examined.

Although the conversion factors obtained for the various shapes are reliable, and well defined, their use has been infrequent, largely due to the fact that some assumption has to be made concerning grain shape. This has led to some confusion as different authors have referred to the "grain size", without referring to the measure used, i.e. mean linear intercept, maximum linear intercept, or equivalent diameter, or other measures related to the plane of polish. It would appear reasonable to describe the "grain size" in terms that are easily measured, and which do not require any assumptions in their derivation. These measures are described in the following section.

2.3 EXPERIMENTAL MEASUREMENT OF GRAIN SIZE

There are three commonly used methods of grain size measurement which cover both quality control and research activities. These are –

(i) Comparator Chart Method

(ii) Heyn Intercept Method

(iii) Jeffries' Planimetric Method

Full details of these methods are described in the ASTM Standard E112, but the essential outlines are described here.

2.3.1 The Comparator Chart Method

A metallographic specimen is prepared and etched to reveal the relevant grain structures. The specimen is examined at a known magnification, usually x100 diameters, and the structure is compared with a series of charts showing a range of grain sizes. Each of the grain sizes shown in the charts has an associated ASTM grain size number, and the grain size number most closely describing the grain size in the test specimen is quoted. The ASTM grain size number, G, has a mathematical basis defined by –

$$n = 2^{(G-1)} \qquad \text{(equ. 2.38)}$$

where n is the number of grains per square inch as seen at a linear magnification of x100 diameters. Thus, for an ASTM grain size number of G = 1, there is one grain per square inch at the specified magnification of x100 diameters, or 10,000 (i.e. 100 x 100) grains per square inch at a magnification of x1. For each unit increase in the ASTM grain size number, the number of grains per square inch is doubled. Thus with an ASTM grain size number of 8, there are 2^7 or 128 times as many grains per square inch as are found at an ASTM grain size number of 1 i.e. 1.28 million grains per square inch at a magnification of x1 (i.e. real size).

The comparator method gives a relatively coarse measure of grain size, each unit of the ASTM scale representing a change of about 40% in grain diameter. The accuracy of the method is estimated to be about + or – 0.5 for G, thus giving an estimate of grain diameter within 20% for the real value. This method has proved to be extremely useful for the assessment of the prior austenite grain size in transformable carbon and low alloy steels, because the grain boundaries of the high temperature phase, austenite, are rendered less distinct by the transformation to other stable or metastable phases at low temperatures. The McQuaid Ehn grain size test involves carburising at a specific temperature in the austenite range until the carbon content is hyper-eutectoid, and the prior austenite grain boundaries are decorated with cementite upon cooling. The prior austenite grain size is clearly revealed by the obvious continuous network of cementite (Fe_3C). This test is used to distinguish grain-refined steels (ASTM No. 7 or higher) from non-grain-refined steels, but cannot be said to describe the austenite grain size under conditions other than those imposed by the test (8hrs. at 927°C in a carburising pack). The austenite grain size at various stages in the hot deformation, or in other processing stages, can be estimated on quenched samples when the martensitic or bainitic transformations may take place, and as martensite of bainite plates are confined to a specific grain, the prior austenite grain structure can still be seen. Methods of revealing these boundaries are described in some detail in the ASTM Standard.

Several different types of chart are available, covering geometrically drawn grids, microstructures having cementite delineated prior austenite grains, or transformed polygonal ferrite, in order to ease the comparisons. Because of the mathematical basis of the ASTM grain size number, very coarse grains (coarser than ASTM 0) or very fine grains (finer than ASTM 10) can be assessed using different known magnifications and making subsequent arithmetical corrections. These and other aspects are dealt with in the Standard.

2.3.2 The Heyn Intercept Method

The Heyn intercept method is probably the most commonly used method for grain sizes in non-transformable alloys and for the low-temperature stable phases of transformable alloys, because the grain boundaries can be clearly defined by etching procedures. In this method, a linear intercept is made either by drawing a line on a micrograph, or by traversing a metallographic specimen under the crosswire in the optical microscope. Given the real length of the traverse, and the number of grains intercepted, the number of grains per unit length, or it's reciprocal, the mean intercept length (Heyn intercept) are easily calculated. The ASTM Standard E112 recommends a 500mm intercept length with the magnification adjusted to give about 50-200 grains in this length.

In a single phase alloy, the intercept length is started within a grain and will (presumably) end within a different grain. The number of grains (including the two half grains at the ends) will then be equal to the number of grain boundaries intersected, provided that there is only one phase. In a two-phase α-alloy, the total length of the intercept will be distributed between the two phases as described by equation 2.17, i.e. for a total length of intercept, L, the length of intercept in the α-phase, L_α, is given by Lf_α. The number of α-grains intersected, n_α, is then used to calculate the mean linear intercept of the α-grains, i_α,

$$i_\alpha = Lf_\alpha/n_\alpha \qquad \text{(equ 2.39)}$$

The number of β-grains can be counted in the same traverse and used to determine the mean linear intercept of the β-grains.

$$i_\beta = L(1 - f_\alpha)/n_\beta \qquad \text{(equ. 2.40)}$$

This principle can also be applied to alloys with more than two phases.

The statistical errors associated with the measurement of the mean linear intercept were considered by J. H. Woodhead (1967), see Blank and Gladman (1970). From the frequency distributions of Hull and Houk, he suggested that the ratio of the standard deviation of the individual intercept lengths σ_i, to the mean intercept length, i, (i.e. the relative error) was virtually constant for a variety of regular and space filling polyhedra, the value of σ_i/i being about 0.7. Thus we may write –

$$\sigma_i/i = k = 0.7 \qquad \text{(equ. 2.41)}$$

Now the standard error of the mean value of the intercept length, SE_i, based on a number of grains, n, is given by standard statistical treatment as –

$$SE_i = \sigma_i/n^{1/2} \qquad \text{(equ. 2.42)}$$

If we express the standard error of the mean as a fraction of the mean value, we get a relative error of the mean, RE_i, i.e.

$$RE_i = SE_i/i \qquad \text{(equ. 2.43)}$$

If a specific relative error of the mean is required, e.g. a relative error of 0.01 (or 1% of the average), the 95% confidence limits will be i $(1-2RE_\alpha)$ and i $(1+2RE_\alpha)$. The value of 2 appearing in these limits is an approximation of the appropriate value of the statistic, t, when n is large (see Fisher and Yates 1938). If the 95% confidence limits need to be i - 0.01i and i + 0.01i, then the relative error would have to be $RE_i = 0.005$. Rearranging Equations 2.41 –2.43, gives –

$$RE_i = SE_i/i = (\sigma_i/i)/n^{1/2} = k/n^{1/2} \qquad \text{(equ. 2.44)}$$

With a value of $RE_i = 0.005$, and Woodhead's suggested value of k = 0.7, then

$$n = (0.7/0.005)^2 = 19{,}600$$

A re-examination of the frequency distributions for the tetrakaidecahedra for both areal interceptions (Hull and Houk, 1953), and linear intercepts (Gladman, 1997), suggest a lower value of k, of about 0.5. Thus, equation 2.44, can be written as –

$$RE_i = 1/(2n^{1/2}) \qquad \text{(equ. 2.45)}$$

The substitution of k = 0.5, rather than Woodhead's value of k = 0.7, roughly halves the number of intercepts required for a given relative error and confidence level.

2.3.3 Jeffries Planimetric Method

This method is based on the assessment of the number of grains appearing in a planar section of some 5000mm² in area. The magnification is adjusted so that about 50 grains are counted. Grains intersected by the perimeter of the area are counted as half grains and are added to the number of grains contained entirely within the area. The grain size is expressed as the number of grains per square millimetre, n_A, and is given by

$$n_A = NM^2/5000 \qquad \text{(equ 2.46)}$$

where N is the number of grains in an area of 5000 mm²
and M is the linear magnification.

Evaluation of the statistical errors associated with this type of measurement can be carried out by the standard statistical procedure of repeating the assessment a number of times on different fields. If F fields are counted, each yielding a result N_f, then the standard deviation σ_N of N_f is obtained from –

$$\sigma_N = [(\Sigma(N_f)^2 - (\Sigma N_f)^2/F)/(F-1)]^{1/2} \qquad \text{(equ. 2.47)}$$

The corresponding standard deviation for n_A can be obtained simply by multiplying σ_N by $M^2/5000$. Alternatively, applying the method suggested by Woodhead, 1967, for the mean linear intercept to the mean areal intercept, a, the equation for the statistical error of the mean areal intercept is virtually identical to Equation 2.45.

$$RE_a = 1/2n^{1/2} \qquad \text{(equ. 2.48)}$$

where n is now the number of grains per square millimetre.

The 95% confidence limits of the mean area of intersection can be obtained directly from Table 2.7, and the reciprocal of these limits gives the corresponding limits of the number of grains per unit area. Although the ASTM Standard recommends the counting of 50 grains, equation 2.17 would yield a relative error of the mean of 0.07, with relatively

Table 2.7 Accuracy of the Mean Linear Intercept, i.

Relative Error the Mean RE_i	Number of Grain Intercepts Required n	Lower 95% Confidence Limit	Upper 95% Confidence Limit
0.005	10,000	0.99i	1.01i
0.010	2,500	0.98i	1.02i
0.025	400	0.95i	1.05i
0.050	100	0.90i	1.10i

wide 95% confidence limits of 0.86a and 1.14a. The corresponding 95% confidence limits of the number of grains per unit area would be the reciprocals i.e. 1/0.86a or 1.16a, and 1/1.14a or 0.88/a. Increasing the number of grains counted to about 100 would give limits of 0.90i and 1.10i as shown in Table 2.7.

2.3.4 Relationships between Grain Size Measures

There are some measures of grain size that are mathematically related to other measures. For example the Jeffries Planimetric measure is mathematically related to the ASTM number using equations 2.38 and 2.46 and correcting for the magnifications and units, so that

$$n_A = 15.5 \times 2^{G-1}$$

where n_A is Jeffries number of grains per mm^2
and G is the ASTM Grain Size number.

The mean linear intercept is also mathematically related to the grain boundary area per unit volume through equation 2.26. However, most other relationships rely on some assumption about the grain shape, or require some change of definition. For example, one description of the grain size is the nominal diameter. The nominal diameter is defined as the diameter of a circle having the same area as the average grain intercept area obtained by the Jeffries method. Thus the nominal diameter is a derivative of the Jeffries Number and the ASTM grain size number. It should be noted that the nominal diameter is related to the sectional area and is not derived from the 3D concept of a sphere having the same volume as the average grain volume. Full details of approximate relationships between grains per unit volume, grains per unit area on a plane section, mean linear intercept, and the ASTM Grain Size Number, are fully described in ASTM E112, and have been discussed elsewhere (Gladman, 1997). The correspondence between the Jeffries, Heyn Intercept and ASTM No. measures of grain size are shown in Table 2.8. It should be pointed out that the mean linear intercept offered in the ASTM Tables is defined as the mean linear intercept of a circle of nominal diameter. The mean linear intercept of a shape of given area, however, is shape sensitive and would be different according to the shape selected. Nevertheless the mean linear intercept obtained for an ASTM grain size of G = 2 (Jeffries count of 31mm^{-2}), is given as 165μm from the ASTM conversion based on the intercept of a circular grain, and would only change to 152 microns for hexagonal grains of equivalent area, or 160μm for mixed square and octagonal grains of the same average area. Using the data given in Tables 2.4 – 2.6, the mean linear intercepts obtained for ASTM2, based on the assumption of 3D tetrakaidecahedral grains would be 154 microns and for 3D cubes would be 147 microns. Although the mean linear intercepts corresponding to given ASTM numbers do depend on

Table 2.8 Correspondence Between Grain Size Measures.

ASTM Grain Size Number	Nominal Diameter d_n µm	Jeffries Method No./mm^2	Mean Linear Intercept µm	
			ASTM Spheres	TKDH
0	360	7.8	318	311
1	254	15.5	225	220
2	180	31	159	155
3	127	62	113	110
4	90	124	80	78
5	64	248	56	55
6	45	496	40	39
7	32	992	28	27
8	22	1984	20	19
9	16	3968	14	14
10	11	7936	10	10

the assumed shapes, the dependence is generally small and the ASTM tables give an adequate guide to the correspondence between the various grain size measurements. The differences between the values obtained by the ASTM conversion and those obtained using 3D tetrakaidecahedra are illustrated in Table 2.8, and in all cases are less than 5%.

REFERENCES

Aboav D., 1992. Materials Science Forum, vols. 94-96, p.275.

Andrews K. W., 1973. Physical Metallurgy – Techniques and Applications. Geo. Allen and Unwin Ltd., London

ASTM Standard E112.

Barratt C. S., 1952. Structure of Metals (2nd Ed.). McGraw Hill.

Blank J. R. and Gladman T., 1970. Tools and Techniques in Physical Metallurgy. Ed. F. Weinberg, Marcel Dekker Inc. New York.

Delesse M., 1848. Annaler des Mines, vol.13, p. 379.

Desch C. H., 1919. Journ Inst Metals, vol.22, p.241

Fisher R. A. and Yates F., 1938. Statistical Tables. Oliver and Boyd Ltd., London.

Gladman T. and Woodhead J. H., 1960. Journ Iron and Steel Inst., vol.194, p.189.

Gladman T., 1963. Journ Iron and Steel Inst., vol.201, p.1044.

Gladman T., 1997. The Physical Metallurgy of Micro-alloyed Steels. Inst. Of Materials, (Book 615), London.

Grignon P., 1776. Memoir de Physique, Paris.

Hilliard J. E. and Cahn J. W., 1961. Trans AIME, vol.221, p.344.

Hull F. C. and Houk W. J., 1953. Trans AIME, (Metals Division), vol.197, p.565.

Kelvin, 1887. Phil. Mag. Vol.24, p.503.

Kelvin, 1894. Proc. Roy. Soc. vol.55 p.1.

Malinsky M., Cwajna J. and Chapronski J., 1992. Materials Science Forum, vol.94-96. p.285.

Matzke E. B., 1943. American Journ of Botany, vol.33, p.58.

Matzke E. B. and Nestler J., 1943. American Journ of Botany, vol.33, p.130.

Rade L. and Westergren B., 1992. Beta Mathematics Handbook. Studentlitteratur, Lind.

Rosiwal A., 1903. Bull. Geol. Soc. Am. vol.14, p.466.

Rutherford J. J. B., Arborn R. H. and Bain E. C., 1937. Metals and Alloys (December), p.345.

Shuttleworth R., 1952. Imperfections in nearly perfect Crystals, W. Shockley ed., Wiley, London p.343.

Smith C. S., 1952. Metal Interfaces. ASM, Cleveland, p.65.

Tomkeieff, 1945. – see Underwood, 1970.

Underwood E. E., 1970. Quantitative Stereology, Addison-Wesley Publishing Co., Reading, Mass.

Williams W. M. and Smith C. S., 1953. Trans Amer. Soc. Metals, vol.45, p.533.

Woodhead J. H., 1967. Unpublished Work.

Chapter 2 Exercises

1. Draw the Schlegel diagram for

   (a) a cube.

   (b) an octahedron.

2. (a) What is the mean area of the grains in a planar section of a material exhibiting an ASTM Grain Size of ASTM 6 ?

   (b) What is the Jeffries Number for this grain size?

3. (a) What is the standard deviation associated with a volume fraction of 0.5, determined by point counting, using 400 points?

   (b) Using lineal analysis, how many α-particles need to be intersected to give this same standard deviation?

CHAPTER 3

Grain Growth

3.1 ESTABLISHING A GRANULAR STRUCTURE

Any change in the microstructure of a material involves the movement of grain boundaries, but may also involve changes other than simple grain growth. Three of the classical ways of changing the microstructure to develop a new granular array are solidification, transformation in allotropic materials, and primary re-crystallisation of deformed materials. In all of these processes, the driving force derives from energy changes that are not related to grain boundary energy. The formation of a granular array by solidification and by allotropic transformation is brought about by changes in the chemical free energy due to the phase changes. Because of this, the grain structure is initially dependent upon the nucleation and growth rates associated with the degree of under-cooling, and therefore with factors such as the cooling rate or isothermal transformation temperature. Nucleation and growth transformations of this kind involve changes in chemical free energy, surface energy (of the growing nucleus), and strain energy, (associated with volume changes during transformations). The chemical free energy provides the driving force, whilst the surface energy and strain energy terms are seen as forces opposing the change in structure. Primary re-crystallisation also shows this same characteristic, but in this process, the driving force is associated with the elimination of dislocations and their associated strain energies.

The under-cooling effect for transformations is such as to encourage nucleation when the under-cooling is pronounced, and the lower temperatures associated with pronounced under-cooling retard the growth rate of activated nuclei. Thus the use of a high degree of under-cooling will give a finer grained structure than is produced when transformation occurs with a lower degree of under-cooling. The situation is slightly different for transformations that occur on heating, where increasing the superheat will increase the nucleation rate but the higher temperature will also increase the growth rate. Nevertheless, the increased nucleation rate effect should dominate at high superheat levels. The microstructure formed after solidification or transformation is a result of the combined effects of nucleation and growth, and, because the nucleation events are highly unlikely to occur with any regularity, the grain structure will contain grains of differing shapes brought about by collision of growing grains. The transformed (or solidified) structure with its grains of different shapes and sizes then provide the system in which grain growth can occur. The grain growth process is driven entirely by the reduction in interfacial area that occurs as certain grains grow and consume their neighbours. It is a thermally activated process, which proceeds at a rate that is controlled by temperature.

Using this as a definition of grain growth poses a question about sub-grain structures. Sub-grains having low energy boundaries, and with very low misorientations between most of their neighbouring sub-grains, are formed by a recovery treatment of cold worked materials. The retained dislocations re-arrange themselves into a lower energy array of sub-grain boundaries. Primary re-crystallisation results in the removal of the sub-grain structure, and again produces a re-crystallised structure as discussed earlier. The elimination of the low energy sub-grain boundaries may be regarded as a sub-grain growth process, and, as such, will be considered in a later section. In this section, grain growth will be considered in the sense of the definition given above – a process of certain grains growing and consuming their neighbours, and, in so doing, reducing the grain boundary energy.

3.2 NORMAL GRAIN GROWTH

Grain growth occurs with an accompanying decrease in grain boundary area and consequently a reduction in the energy associated with grain boundaries. The driving force for grain growth derives solely from the reduction in grain boundary energy of the system, and does not involve phase transformation.

The irregularities in the microstructure following solidification, transformation, or re-crystallisation, resulting from the precise location of grain nuclei and the accidents of growth and collision, will provide the means whereby grains can grow, but the fundamental force derives from the reduction in energy associated with larger grains. As described in Chapter, 2, the grain boundary area per unit volume can be derived for grains of various shapes and is expressed in units of mm^2/mm^3 or mm^{-1}. For example, taking data for the area and volume of the tetrakaidecahedron (Rutherford, Arborn and Bain, 1937), yields –

$$A = (6 + 12\sqrt{3})e^2 \qquad \text{(equ. 3.1)}$$

$$V = 8\sqrt{2}.e^3 \qquad \text{(equ. 3.2)}$$

where A is the surface area of the tetrakaidecahedron
V is the volume
and e is the edge length.

The area per unit volume for this shape is –

$$A/V = (6 + 12\sqrt{3})/8\sqrt{2}.e \qquad \text{(equ. 3.3)}$$

If the polyhedra are stacked to fill space as shown in Fig. 2.3, the interfacial area per unit volume will be one half of that shown in equ. 3.3, because each face of the tetrakaidecahedron is shared by two grains, so that the area per unit volume in the array, A_v, is given by –

$$A_v = (6 + 12\sqrt{3})/16\sqrt{2}.e \qquad \text{(equ. 3.4)}$$

This clearly shows that as the tetrakaidecahedral grains increase in size, the grain boundary area per unit volume decreases, (Ewing and Rosenhain, 1900), and is reciprocally related to the edge length of the tetrakaidecahedron, e, or indeed to any linear measure of the size of the tetrakaidecahedron.

Not all grains can grow, as there is the slight matter of conservation of mass! Only certain grains can grow and therefore other grains must shrink. Further, if some grains grow whilst other shrink, the mean grain volume will remain the same, so that an important feature of grain growth is the disappearance of certain grains. Thus, some grains grow into other grains and ultimately consume them. It is this feature that has led to grain growth being described as a cannibalistic process. The number of grains per unit volume decreases as a result of the grain growth process.

Grain growth can also be described as a tactile process. Each grain will be aware (if we can personalise the grains), of the standing of it's immediate neighbours. Not only must the topological rules be obeyed, but also the surface energy considerations will be important in deciding whether a grain should grow or shrink. For simplicity, let us assume that all grain boundary energies are equal. If a grain has an edge where the angle between its two adjacent faces is less than 120°, then the edge will move inwards and the grain will shrink. If a grain has a corner where the angles between it's three associated edges are less than 109°28', then the corner will move towards the grain centre and the grain will shrink. The converse is also true. Grains with fewer faces will be those with smaller angles between their adjacent faces and between adjacent edges at their corners, and will tend to shrink. Conversely, grains with more faces will grow. All of this is in accord with the general observation that large grains consume smaller grains until they disappear and the growing grain meets a new neighbour, which in turn may be larger or smaller than itself.

There have been many approaches to the theory of grain growth based on the above ideas, made with a view to predicting the kinetics of growth and the evolution of the grain size. The nature of the problem has been quite succinctly stated by Hunderi and Ryum, 1992 –

'To do this (predict grain coarsening) one has to describe the growth and shrinkage of every grain in the assembly'.

3.2.1 Analytical treatments

Early analytical solutions, such as that of Hillert (1965), described a situation in which the rate of growth (or shrinkage) of a grain was dependent upon the size of the grain relative to some critical size.

$$dR/dt = \alpha(1/R_c - 1/R) \qquad \text{(equ. 3.5)}$$

Grains having radii, R, larger than the critical size, R_c, would grow, whilst grains smaller than the critical size would shrink. The rate of growth of a grain, dR/dt, was thus ascribed to its size and would be independent of the size of its neighbours. Hunderi and Ryum objected to this assumption and cite the Weaire – Aboav Law, (Aboav, 1970, Weaire, 1974), which suggests that large and small grains are not distributed randomly. However, as Hillert did not precisely specify the physical significance of his 'critical size' a priori, his theory would be satisfactory if all larger grains were surrounded by smaller than average grains, but unfortunately would be less satisfactory if the large grains were surrounded by larger than average grains, whilst others of a similar size were surrounded by average sized grains or by smaller than average grains. Despite the assumption of an independence of spatial position, Hillert's analysis of the grain growth process was useful in providing an insight into the growth rates and the evolution of the granular structure. Hillert's approach led to a parabolic rate of growth with –

$$R^2 - R_i^2 = kt \qquad \text{(equ. 3.6)}$$

where R is the grain radius after a time, t,

and R_i is the initial grain radius.

This equation had also been derived by Burke and Turnbull, (1952), based on the concept of local grain boundary curvature as the driving force. There have been many experimental assessments of the growth exponent in equation 3.6, many of which differ

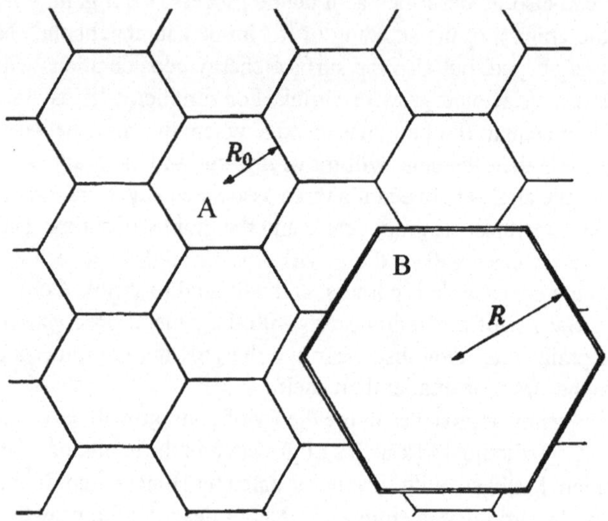

Fig. 3.1 Two Dimensional Representation of the Growth of a Large Tetrakaidecahedron in a Matrix of Smaller Ones.

from the value of n = 2. Some results on high purity materials show the expected value of n = 2, but many results show higher values of n. The reasons for this are not obvious, but there are two reasons that are often cited.

(i) The assumption is often made that R_i may be small with respect to R, so that the equation 3.6 could be re-written as –

$$R = kt^{1/n} \qquad \text{(equ. 3.7)}$$

At short times when R_i cannot be ignored, significant deviations from n = 2 would be expected.

(ii) At longer times, or coarser grain sizes, the driving force for grain growth decreases, and, as will be shown in a later section, growth inhibiting forces begin to play a more important role. In fact, many authors have referred to a 'limiting grain size' above which grain growth will not occur.

An analytical model was also developed to describe the driving forces for the growth of individual grains (Gladman, 1966). Here, the energy changes associated with particular grains when the grains either grow or shrank were obtained by recognising that, as one grain grows, it's own surface area will expand and there will be an increase in energy. However, the growth of this grain will consume neighbouring grains and, more importantly will eliminate grain boundary area that only belongs to the neighbouring grains. It was then assumed that the energy density associated with grain boundaries between pairs of neighbours would be inversely proportional to some linear dimension of the neighbour's size as shown by equation 3.4. The assumptions made were –

(a) That all neighbours have the same size as one another.

(b) That their grain boundary energy is more or less uniformly distributed through the volume.

If we allow the growing grain to have a size, R, which is not the same as that of the neighbouring grains (radius R_o), as depicted in Fig. 3.1, the rates of change of the grain boundary area, and therefore energy, can be derived as follows.

Let a tetrakaidecahedral grain of radius R (the radius being one half of the distance between opposite hexagonal faces) grow by a small amount, δR. The surface area of the grain is given by –

$$A_1 = (6 + 12\sqrt{3})\,e^2 \qquad \text{(equ. 3.8)}$$

where e is the edge length of the tetrakaidecahedron, and the radius, R, is given by

$$R = \sqrt{6}.e/2 \qquad \text{(equ. 3.9)}$$

Substituting for e in Equation 3.8 gives –

$$A_1 = (4 + 8\sqrt{3})\,R^2$$

So that

$$A_1 + \delta A_1 = (4 + 8\sqrt{3})(R + \delta R)^2$$

and by standard differential calculus –

$$dA_1/dR = 2(4 + 8\sqrt{3})R \qquad \text{(equ. 3.10)}$$

If the area change, dA_1/dR, is converted to an energy change, dE_1/dR, by including the surface tension, γ, then –

$$dE_1/dR = 2(4 + 8\sqrt{3})\,R\gamma \qquad \text{(equ. 3.11)}$$

If this force (force in the sense of being a rate of change of energy), is then expressed as the rate of change of energy per unit area of the advancing interface, then –

$$(dE_1/dR)/A = (dE_1/dR)/(4 + 8\sqrt{3})R^2 = +2\gamma/R \qquad \text{(equ. 3.12)}$$

Because the energy change is positive when the grain grows, it is not a driving force for grain growth, but a retarding influence. The rate of change of energy per unit area of advancing interface expressed in equation 3.12, is independent of the shape chosen. This is the value used by Zener, (Zener 1949), as the driving force for grain boundary movement in his signal work on particle drag. Because of the adverse sign, Zener's use of Equation 3.12 as a driving force could only have referred to a shrinking grain, whether spherical or otherwise.

In the case of the growing grain, however, the adverse effect of the boundary expansion has to be overcome by elimination of the boundaries between neighbouring grains. Using the energy density approach, the area of grain boundary per unit volume, A_v, of an array of tetrakaidecahedra of radius R_o, is given by –

$$A_v = A/2V \qquad \text{(equ. 3.13)}$$

Substitution from Equations 3.4 and 3.9 gives –

$$A_v = (4 + 8\sqrt{3})\,R_o^{\,2}/(2.8.\sqrt{2}\,(2R_o/\sqrt{6})^3) \qquad \text{(equ. 3.14)}$$

Thus if the large grain grows and increases it's own volume, it will eliminate A_v units of grain boundary area per unit volume from neighbouring grains in the matrix consumed.

When the growing tetrakaidecahedral grain of radius R and volume V grows to a radius R + δR, and volume V + δV, the volume increase is –

$$\delta V = 3R^2\,(8\sqrt{2}\,(2/\sqrt{6})^3)\,\delta R \qquad \text{(equ. 3.15)}$$

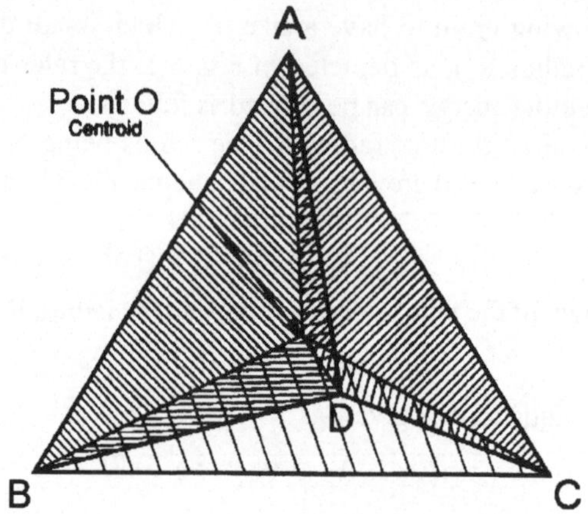

Fig. 3.2 The Shrinking Tetrahedron.

and the area of grain boundaries eliminated, δA_2, in the consumed matrix is –

$$\delta A_2 = A_v \delta R = A_v\ 3R^2\delta R \qquad \text{(equ. 3.16)}$$

$$= (4 + 8 .\sqrt{3})\ 3R^2\delta R/2R_o \qquad \text{(equ. 3.17)}$$

Further, if this area eliminated (note it can be regarded as a driving force as Equation 3.17 represents a loss of grain boundary area, and therefore a reduction in energy) is then expressed in terms of energy per unit surface area of the growing grain, then the rate of change of energy of the boundaries eliminated, dE_2/dR, is given by –

$$dE_2/dR = -3\gamma/2R_o \qquad \text{(equ. 3.18)}$$

Thus the total rate of energy change per unit area of growing grain boundary, dE/dR, is the sum of the Equations 3.12 and 3.18.

$$dE/dR = dE_1/dR + dE_2/dR$$

$$dE/dR = (2/R - 3/2R_o)\gamma \qquad \text{(equ. 3.19)}$$

Thus the driving force for growth of the grain of radius R is not only dependent upon its radius R, but is also dependent upon the size of its immediate neighbours (R_o). An important aspect of equation 3.19, is that the grain of radius R, will only undergo robust growth when the rate of energy change is negative, i.e. –

$$(2/R - 3/2R_o) < O$$

$$\text{or } 2/R < 3/2R_o$$

$$R > 4R_o/3 \qquad \text{(equ. 3.20)}$$

In other words, a grain must have a size advantage (R/R_o) of at least 4/3, in order to reduce the energy of the system by growing. If equation 3.19 is rewritten to express the size of the growing grain in terms of the size advantage, $Z = R/R_o$, (Gladman, 1966), then the driving force, G, is given by –

$$G = -dE/dR = -(2/Z - 3/2)\ \gamma/R_o \qquad \text{(equ. 3.21)}$$

If the size advantage, Z, is less than 4/3, the energy change will be positive, and growth will not be favoured. If the size advantage is greater than 4/3, the energy change will be negative and growth can occur. The rate of change of energy when growth is favoured will increase as the size advantage increases (2/Z decreases), and as the size of neighbouring grains decrease at a given size advantage. Although equation 3.21 was not used to predict the evolution of the grain structure with time and temperature, but was mainly used to understand growth inhibition, (Gladman, 1966), it could be used to to follow the development of the grain structure, as in the Hillert model. The general formulation for the rate of grain growth is given by –

$$dR/dt = MG = -M\gamma(2/Z - 3/2)/R_o \qquad \text{(equ. 3.22)}$$

where M is the grain boundary mobility,

γ is the grain boundary energy,

Z is the size advantage (R/R_o),

and R_o is the radius of neighbouring grains.

The evolution of the grain structure, however, could be quite complex, in that the growth of a grain with a sufficient size advantage will increase its size advantage over existing neighbours, i.e. Z increases with time. As growth continues, the neighbouring grains will be completely consumed until the growing grain ultimately collides with other growing grains, thereby decreasing its size advantage, Z, and also increasing the size of neighbouring grains, R_o, both of which will at least slow the growth rate. In fact, the collision of growing grains creates a completely new environment for each of the grains, not all of which can continue to grow. This complexity is exactly reflected in the quotation from Ryum and Hunderi concerning the 'growth and shrinkage of every grain in the assembly'.

Although Equation 3.21 was developed as a 'mean field' theory and is capable of being tested as such, it can also be treated as a specific geometric model in the sense that, at some part of the structure, there may be a grain of radius R surrounded by other grains of radius R_o. From this point of view, it is possible to examine other geometric arrangements, such as the shrinking tetrahedron (Gladman, 1980). It follows that if a grain is much smaller than its neighbours, then the small grain will have a small number of neighbours, and therefore a small number of faces. If a microstructure is formed by the growth of roughly spherical grains that collide (by either a primary re-crystallisation or a transformation process), then 2 or 3 sided lenticular grains with curved faces are unlikely to be formed. Under these circumstances, the smallest grains are more likely to take the form of tetrahedra, a shape that will conveniently fit between four neighbouring grains, and which will, on collapsing, leave a four-rayed grain corner, Fig. 3.2. The area changes involved when the tetrahedron collapses, include the disappearance of the four original triangular faces of the tetrahedron such as that marked ABC in Fig. 3.2, and the extension of the six faces between neighbouring grains at the six edges of the tetrahedron such as that marked AOB. Whilst accepting that any four-sided grain may not be a regular tetrahedron, the mathematics of the area changes are much easier for the regular tetrahedron where the shrinking process is achieved by the simultaneous movement of the four surfaces in directions normal to those surfaces, resulting in their disappearance at the centre of gravity (point O in Fig. 3.2).

Let the shrinkage process be defined in terms of the change of the 'inradius' (i.e. the radius of an inscribed sphere), which describes the movement of the surfaces in a normal direction

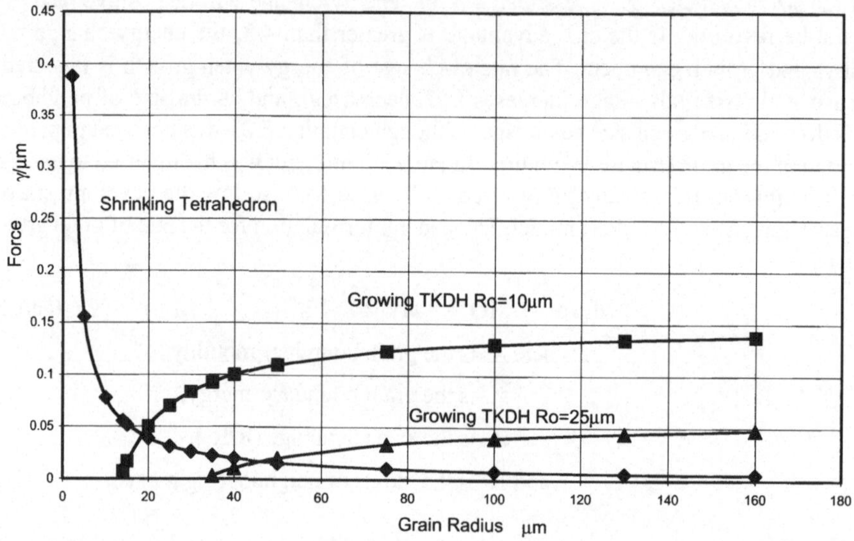

Fig. 3.3 Driving Force for Grain Growth.

towards the centre of gravity. The inradius, R_i, is related to the edge length 'a', (Rade and Westergren, 1992), by –

$$R_i = a\sqrt{6}/12 \quad \text{(equ. 3.23)}$$

The combined surfaces of the tetrahedron that are eventually eliminated by grain shrinkage have a total surface area, S_1, given by (Fig. 2.11) –

$$S_1 = \sqrt{3}\,a^2 = 24\sqrt{3}\,R_i^2 \quad \text{(equ. 3.24)}$$

The six surfaces forming the interfaces that are created as the tetrahedron disappears (i.e. the triangular interfaces such as ΔAOB in Fig. 3.2) have a combined surface area, S_2, given by –

$$S_2 = 6\sqrt{2}\,a^2/8 = 18\sqrt{2}\,R_i^2 \quad \text{(equ. 3.25)}$$

If we differentiate Equations 3.24 and 3.25 with respect to R_i, and sum the rates of change of area, bearing in mind that S_1, increases with R_i whilst S_2 is disappearing, we obtain the rate of change of area, S, as –

$$dS/dR_i = 48\sqrt{3}R_i - 36\sqrt{2}R_i \quad \text{(equ. 3.26)}$$

Converting this to energy and expressing this as the change in energy per unit area of interface, the driving force for growth of the tetrakaidecahedron is given by –

$$dE/dR_i = (2 - \sqrt{(3/2)})\gamma/R_i \quad \text{(equ. 3.27)}$$

This is a positive energy change so that for grain shrinkage, i.e. R_i decreasing, the energy of the system is reduced.

It can be appreciated, from Equation 3.27, that the growth of a tetrahedral grain from a 4-rayed grain corner would result in an increase in the energy of the system, and that it is the disappearance of such grains that lead to a lowering of the energy of the system. Furthermore, from equation 3.27, the driving force per unit area of grain boundary increases as the tetrahedron becomes smaller. It is also interesting to see that the driving force for the shrinking tetrahedron is independent of the size of its neighbours according to equation 3.27, but by implication, the tetrahedron must have only four neighbours, and these neighbours must necessarily be larger than the tetrahedron in order to comply with the surface energy requirements of three grains meeting at an edge and four grains meeting at a corner. Imagine a tetrahedral grain growing from a grain corner in an array of tetrakaidecahedral grains such as that shown in Fig. 2.3. The four grain edges meeting at the corner would approximate the outradius of the tetrahedron. The approximation arises because the tetrakaidecahedron is not a regular polyhedron and therefore the angles between the edges are not equal. Nevertheless, as an approximation, the maximum outradius of the tetrahedron would be defined by the length of the edges of the tetrakaidecahedron and as the outradius is $\surd$ 6/4 times the edge length of the tetrahedron, then the maximum edge length of the tetrahedron is about 1.6 times the edge length of the tetrakaidecahedra. With these dimensions, the maximum volume of a tetrahedron would be less than one twentieth of the volume of each of the tetrakaidecahedra forming the matrix.

From Equations 3.21 and 3.28, it is possible to compare the energy changes associated with the growth of a tetrakaidecahedral grain with those of a shrinking tetrahedron. As the maximum size of a tetrahedron in a matrix that would otherwise be tetrakaidecahedra is governed by the fact that the outradius of the tetrahedron (the radius of a circumscribed sphere), should not exceed the edge length of the tetrakaidecahedron, the inradius of the tetrahedron will never exceed a value greater that $0.2722R_o$, where R_o is defined as the half distance between opposite hexagonal faces of the tetrakaidecahedra, (Gladman, 1966). Thus the minimum driving force for the shrinkage would be –

$$(2 - \surd(3/2)\gamma/0.2722R_o = 2.85\gamma/R_o \qquad \text{(equ. 3.28)}$$

Now the maximum driving force for the growth of a tetrakaidecahedral grain of radius R in a matrix of tetrakaidecahedra occurs when R is infinite, Fig. 3.3, and according to Equation 3.21, is given as $1.5\gamma/R_o$. Thus the minimum driving force for tetrahedral grain shrinkage ($2.85\gamma/R_o$) is always greater than the maximum driving force for tetrakaidecahedral grain growth ($1.5\gamma/R_o$) provided that the maximum size (outradius) of the tetrahedron is governed by the edge lengths in the tetrakaidecahedral matrix, as shown in Fig. 3.3.

Gladman did not expand the work to cover actual grain growth, but more recently some of the more sophisticated features of grain growth have been summarised (Pande, 1992), as follows –

(i) During grain growth, the grain size distribution reaches a steady state that is approximately lognormal.

(ii) The lognormal distribution is developed irrespective of the initial distribution.

(iii) During grain growth, the mean grain size increases whilst the lognormal distribution is maintained.

Some confusion may arise from the use of experimental observations of linear or areal intercept distributions. Methods of correcting the lineal intercept distributions to spatial size distributions (Gladman, 1997), or correcting areal intercept distribution to spatial size distributions, (Hull and Houk, 1992), have been tried (Matsuura and Itoh, 1992), although little has been published on the use of this technique. The distributions of lineal or areal intercepts may not reflect the distribution of spatial measures of size, e.g. volumes, maximum sectional areas, maximum linear intercepts.

3.2.2 Topological Models

Topological approaches to growth have been made, usually based on a two dimensional system. In a two dimensional system (Smith, 1952), the topological equation (equ. 2.1), is modified by the application of the surface energy constraints, i.e. three polygons meet at a corner, two at an edge, and each edge has two corners, each corner having three shared edges. Now let the number of polygons with n corners be P_n.

The total number of polygons P is given by –

$$P = \Sigma P_n \qquad \text{(equ. 3.29)}$$

The number of corners is given by –

$$C = \Sigma n P_n/3 \qquad \text{(equ. 3.30)}$$

and the number of edges if given by –

$$E = \Sigma n P_n/2 \qquad \text{(equ. 3.31)}$$

Substitution into Equation 2.1 yields –

$$\Sigma P_n - \Sigma n P_n/2 + \Sigma n P_n/3 = 1 \qquad \text{(equ. 3.32)}$$

This can be simplified to give the 6 – n rule, i.e. –

$$\Sigma (6 - n) P_n = 6 \qquad \text{(equ. 3.33)}$$

and the average number of sides per polygon,ν, is given by –

$$\nu = 6(1 - 1/\Sigma P_n) \qquad \text{(equ. 3.34)}$$

Thus, in the infinite array, the average number of sides will be six, and any five-sided polygon must be balanced by a seven sided polygon. The consequence of a variation in the number of sides is illustrated in Fig. 2.1. Straight sided polygons with less than 6 sides have internal angles that are less than 120°, whilst for those with more than 6 sides, the angles are more than 120°, as shown in Fig. 2.1(a). Any local adjustment to give triple junctions of 120° will result in convex curvature for polygons with fewer than six sides, and concave curvature for those with more than six sides. Thus, the polygons with fewer than six sides will shrink as the boundaries move towards their centre of curvature, and polygons with more than six sides will grow for the same reason. The polygons with fewer sides will tend to be the smaller grains, whilst those with more sides will generally be the larger grains. The topological argument thus embodies the aspects of angle driven, curvature driven, and relative size effects.

The formulation of the grain growth (or grain shrinkage) law is expressed by the equation (Von Newmann, 1952) –

$$da/dt = \pi M \gamma (n - 6)/3 \qquad \text{(equ. 3.35)}$$

where a is the area of the (2 dimensional) grain,
M is the grain boundary mobility,
γ is the grain boundary energy,
and n is the number of sides of the grain.

The grain boundary mobility, M, defines the boundary velocity per unit of driving force. This equation like many others, also suffers from the defect that all n-sided grains should grow at the same rate, and takes no account of the neighbouring grains.

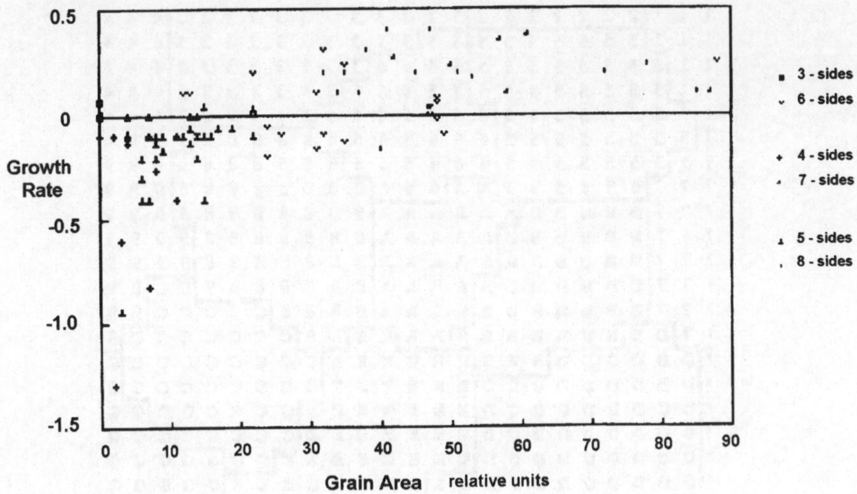

Fig. 3.4 Grain Growth Rates in Succino Nitrile (After Glickman et al).

The problem of neighbouring grains has been addressed (Fradkov, 1992, Heckelmann, Lucke, and Abbruzzese, 1992) using the Weaire-Aboav Law –

$$m_n = a + b/n \qquad \text{(equ. 3.36)}$$

where m_n is the mean number of sides of the polygons adjoining n-sided polygons and a and b are constants.

Fradkov found by experiment, using through thickness columnar grains in aluminium foils that –

$$m_n = 5 + 8/n \qquad \text{(equ. 3.37)}$$

It should be pointed out that m_n is an average value and its use would not in itself overcome the problem of location sensitive growth. This problem has to be dealt with by a statistical approach, (Heckelmann, Lucke, and Abbruzzese, 1992). This statistical model acknowledges that the number of sides of grains neighbouring an n-sided grain will vary, and that the mean number of sides, m_n, of the neighbours of an n-sided grain will also vary.

An experimental study of 2-D grain growth in a succino-nitrile film (Glickman, Kajan, Nordberg, Palmer, Marsh, and Pande, 1992) has provided support for the growth equation (equ. 3.35), as shown in Fig. 3.4. Those grains with less than 6 sides show a negative growth rate so that they shrink and will disappear. Grains with more than 6 sides generally show a positive growth rate. The grains with 6 sides may show small positive growth rates or small negative growth rates, and this may reflect the importance of the precise location of any particular grain with respect to the number of sides. Although the Weaire-Aboav Law shows that the mean number of sides on neighbours of a 6-sided grain is 6.33 (equ. 3.37), there will be some variation thus allowing for growth or shrinkage.

Fig. 3.4, also gives some support to the driving forces expressed by equations 3.21 and 3.28 in section 3.2.1, in that the smaller grains, normally 3 or 4 sides, show large negative growth rates, the larger grains, usually 7 or 8 sided, show modest positive growth rates, whilst the intermediately sized grains may show slow growth or slow shrinkage depending upon location (the nature of its immediate neighbours).

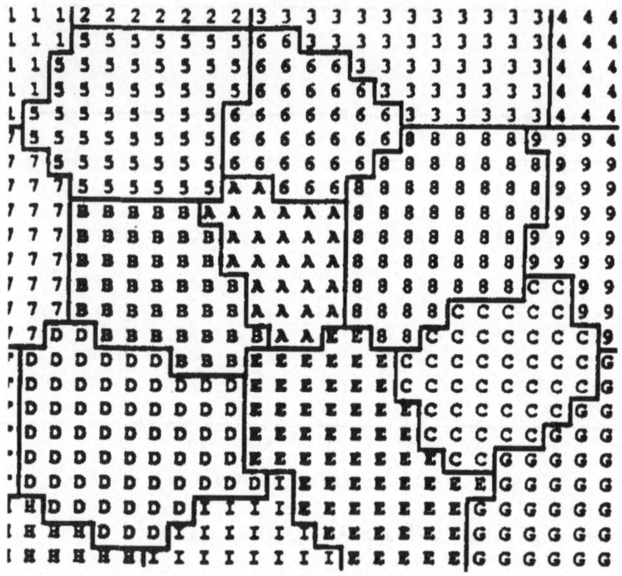

Fig. 3.5 The Pott's Model Showing Spin Orientations and Grain Boundaries in a Computer Model.

In mathematical treatments of grain growth, whether by analytical or topological treatments, the instantaneous rates of change are relatively easy to formulate, as for example in equations 3.22 and 3.35. The main problem arises from the switching that accompanies the disappearance of a small grain. The subsequent growth is then dependent upon the relative sizes of a new set of neighbours. This difficulty has led to computer based approaches where the fortunes of individual grains in an array can be followed and the evolution of the microstructure can be mapped.

3.2.3 Computer Based Models

In the absence of a general analytical solution for the evolution of the grain structure during grain growth, and with the development of computers of ever increasing capability, it was natural that the grain growth process would be modelled by computer methodologies. In the mid-1980's, the 'Exon Group' published several important papers on the computer simulation of grain growth using the Potts model, (Anderson, Srolovitz, Grest, and Sahni, 1983). The Potts model is illustrated in Fig. 3.5 (Ling, Anderson, Grest, and Glazier, 1992). The polycrystalline (2-D) microstructure is discretised with a triangular, (or possibly square) lattice. Each site is given a spin number between 1 and Q that denotes the orientation of the grain in which it is embedded. Here, Q is the maximum number of grain orientations considered. The grain boundary separates regions of different spin numbers, and neighbouring grains have different spin numbers. The growth process is achieved by assuming a random re-orientation at a random site on the lattice and assessing whether this results in an increase or decrease in grain boundary energy. The re-orientation is accepted only if the energy change is not positive. The time span is measured in terms of a Monte Carlo step (MCS), which is equal to the number of re-orientation attempts relative to the number of lattice sites. Ling et al used a copy of a 2-D soap froth as input to the computer and

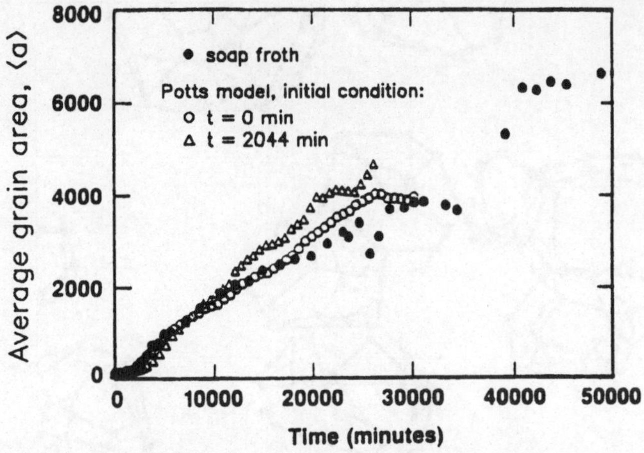

Fig. 3.6 Comparison of Bubble Growth with Computer Predictions in 2-D.

compared the observed bubble growth with the computer prediction. Allowing for the scaling between Monte Carlo steps and time, reasonably good agreement was found, as shown in Fig. 3.6, and the mean area increases approximately linearly with time. Such studies have been extended to include grain size distributions and nearest neighbour side correlation functions, and the models show good agreement with bubble growth in soap froths. Not only were the grain size distributions similar, but also the nearest neighbour side correlations of both the froth and the model, conformed with the Aboav-Weaire Law (equ. 3.37).

The extension of the Potts model to three dimensions has been limited (Anderson, Grest and Srolovitz, 1989), mainly because of the computer time involved. Anderson et al used a 100 x 100 x 100 lattice and followed the grain growth process. The grains developed shapes consistent with an average number of faces of 12.9, i.e. somewhere between the pentagonal dodecahedron and the tetrakaidecahedron, and the time dependence was found to be highly dependent upon the anisotropy. Further work on 3-D systems using the Potts model, although desirable, appears to have been superceded by Vertex models. In the Vertex model, the microstructure is defined by the co-ordinates of the grain corners (whether this is a 2-D or a 3-D model). The corners are regarded as entities that can respond to applied forces and have specified mobilities that represent the proportionality constant between the velocity and the applied force. The forces on each vertex (grain corner) are produced by the changes in total area of the six touching interfaces as the vertex is moved. (In the 2-D model there are only 3 touching edges and the vertex moves in a direction that would minimise the summed length of these boundaries). If all grain boundary energies were equal, the vertices should move so that angles between grain edges approach 120° in the 2-D case, and 109°28' (see Section 2.13) in the 3-D case.

Although the idea of the vertex model was suggested as long ago as 1952, (Fulman, 1952), significant developments of the 3-D model awaited adequate computational facilities, (Kawasaki, Nagai, and Nakaskima, 1987). In a more recent publication Nagai, Ohta, and Kawasaki (1992), have updated their treatment of a 2-D vertex model to include grain boundary curvature effects. The results obtained show the developments of a stable distribution function for the number of sides of the 2-D grains in an array. The calculated rate of growth of grains showed close agreement with the Neumann-Mullins Law –

$$dA/dt = \pi M (n - 6)/3 \qquad \text{(equ. 3.38)}$$

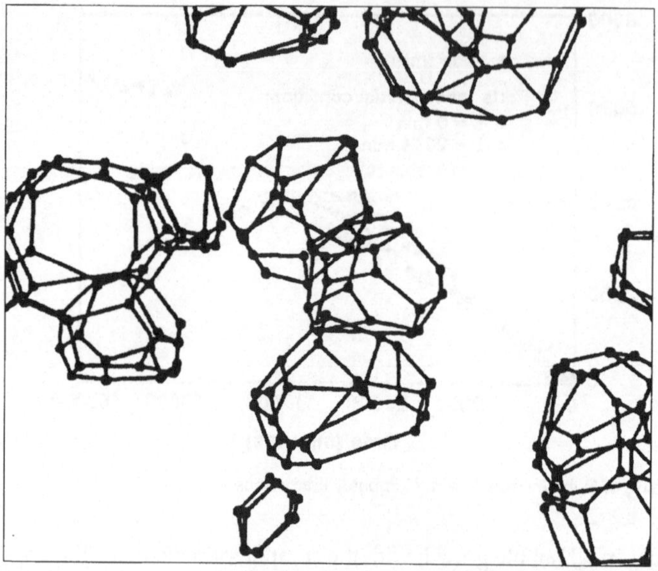

Fig. 3.7 Evolving Cellular Pattern in 3-D Projected onto a Plane, using the Vertex Model. (Only selected cells are shown).

and also with the growth rates for succinonitrile shown in Fig. 3.4.

The early attempts with the 3-D vertex model by these same workers, starting with Voronoi polygons, exhibited a transient grain size distribution function, which ultimately settled into a constant distribution function where only the scaling of the grain sizes differed. The evolving cellular shapes are illustrated in Fig. 3.7, where they are projected onto a plane. Only selected cells are shown. The distribution of the number of faces per cell in the scaling state (where the distribution function is the same) is shown in Fig. 3.8. The average number of faces per cell decreased from about 15.5 to about 13.5 in the scaling state. The results also suggested a 3-D equivalent of the 2-D Aboav-Weaire Law, where the average number of faces per cell $m(n_f)$ of cells adjacent to those having n_f faces is given by –

$$m(n_f) = 12.92 + 29.85 / n_f \qquad \text{(equ. 3.39)}$$

There is little doubt that further advances will be made in the computer modelling of grain growth, and it is possible that 3-D vertex models coupled with energy change rules will provide interesting insights into the growth process. One of the main difficulties remaining, however, is the multiple curvature of grain interfaces due to the non-planar coincidence of the grain corners on a given face.

3.2.4 Other Approaches

Although there have been other approaches, such as the curvature driven grain boundary movement, or angle driven motion, it should have become clear that these concepts have been implied in the analytical, topological, and computer simulation models. None of these pose any significant contradiction, as a large grain in an array will generally be surrounded by smaller grains, it will have more faces than the neighbouring grains, and its grain edge angles

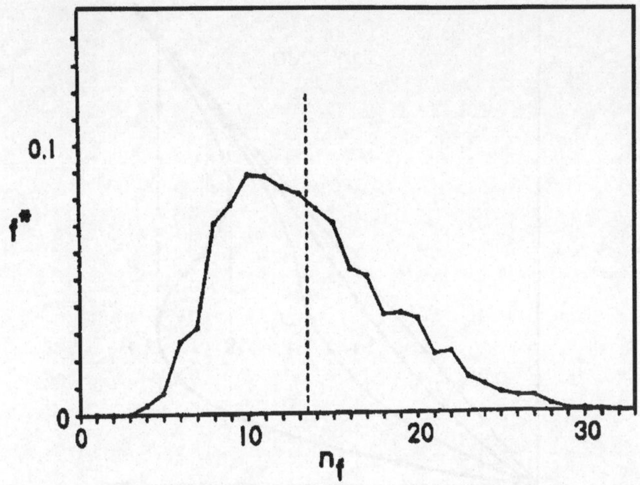

Fig. 3.8 Frequency Distribution of the Number of Faces per Grain in the Steady State Regime of the Vertex Model.

will be larger than 120°, whilst its corner angles will exceed 109°28'. The converse will generally be true for small grains in the array. Although these are general truths, it does not follow that all grains of the same size have the same number of faces, the same edge angle, or the same vertex angle. The precise numbers depend on the characteristics of the neighbouring grains (i.e. vary with location) and this has led to stochastic approaches some of which have already been cited (Pande 1987, 1992). A random statistical term is added to the purely analytical terms to allow for the location sensitive contribution. The main conclusions concerning the mean rates of growth, with respect to cell size, however, are unchanged by the statistical term. It would appear that all of the approaches are based on inter-related features of the growth process. If a grain has few sides, the internal angles are small, and the grain will have convex surfaces as the vertices move to their equilibrium positions. The boundaries will then migrate towards their centres of curvature to minimise the area, thereby providing the conditions necessary for further vertex movement. Variations in the rate of adjustment will increase with increasing deviation from equilibrium, and this will additionally depend upon the neighbouring grains. Thus all of the theories of grain growth provide the same description of the process provided that they correctly account for the constancy of volume of the array, and for the connectivity whereby grains are connected along shared surfaces (Thorvaldsen, 1992).

3.3 ABNORMAL GRAIN GROWTH

In the normal grain growth process, the spread of grain sizes remains relatively small, even though the average grain size may increase dramatically. There is however another form of grain growth, in which a specific grain or grains may grow in a matrix of fine grains that remain virtually unchanged. The reason for this behaviour will be discussed in the next chapter, but there is an important consequence arising from abnormal grain growth. During normal grain growth the growing grains are encroaching on progressively coarser matrix grains and so the grain growth process gradually becomes slower as the driving force is reduced (see Fig. 3.3). On the other hand, with abnormal grain growth, the driving force for

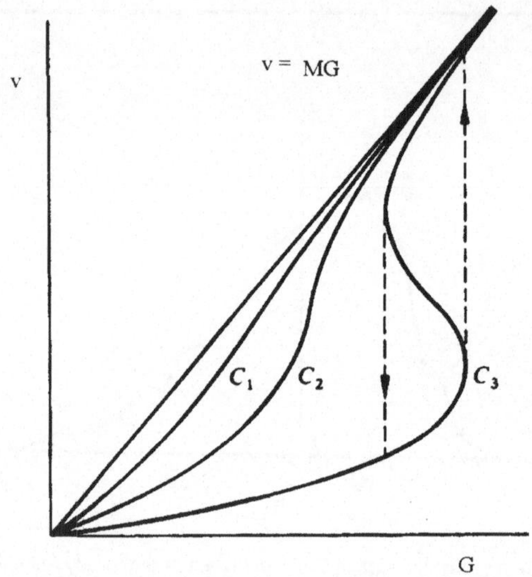

Fig. 3.9 Grain Boundary Velocity, v, as a Function of Driving Force, G, for Three Impurity Levels.

abnormal grain growth increases as the growing grain increases its size advantage over the unchanged matrix grains (also shown in Fig. 3.3). Thus the growth rate of the abnormal grain outstrips those observed in normal grain growth and the abnormal grain size attained may be several orders of magnitude larger than those observed after normal grain growth. This can have very adverse effects on the properties and behaviour of the material. Abnormal grain growth may also be called secondary re-crystallisation and may be used, particularly in materials where coarse grains imparts improved magnetic properties.

3.4 GROWTH VELOCITY AND IMPURITY EFFECTS

Attention was drawn to impurity segregation at grain boundaries (Section 1.6), and early experimental observations had shown that impurities could cause a marked reduction in the rate of boundary migration during re-crystallisation. Lucke and Detert (1975) were the first to try to formulate a quantitative theory for this effect. The boundary velocity v (expressed as dR/dt in equ. 3.22), is given a solute drag term (Lucke and Stuwe, 1963; Cahn, 1962) which effectively reduces the driving force

$$v = M(G - I)$$

where I is the impurity drag force. The magnitude of the drag force depends on the impurity concentration and the boundary velocity. Cahn calculated the value of I from the shape of the interaction energy trough and the variation of the diffusion coefficient of the impurity atoms near the trough, and obtained –

$$G = v/M + \alpha Cv/(1 + \beta^2 v^2) \qquad \text{(equ. 3.40)}$$

where C is the impurity concentration

and α, β are parameters which depend on the shape of the energy trough and the diffusion coefficient profile.

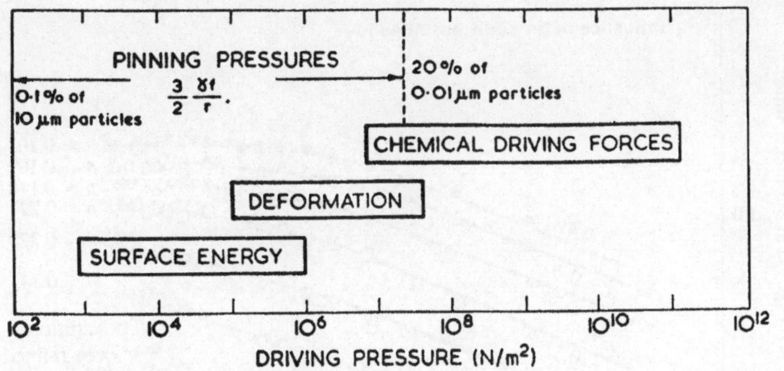

Fig. 3.10 The Driving Force for Microstructural Change. (After Ashby).

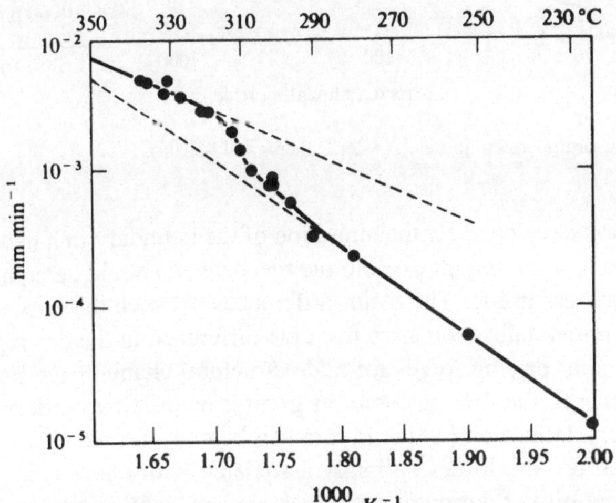

Fig. 3.11 Effect of Iron on Grain Boundary Velocity in Gold. (After Doherty and Martin).

The effect of temperature is implicit because of the temperature dependence of the various parameters (M, α, β, etc). A schematic diagram, Fig. 3.9, shows the complex relationship between the boundary velocity and the driving force when impurities are present in differing amounts. The curves represent three impurity levels, C_1, C_2 and C_3, and at low driving forces, movement of the boundary is accompanied by a repetitive unlocking and relocking of the boundary by the impurity atoms and the velocity decreases as the impurity content is increased from C_1 to C_3. At higher driving forces the unlocking of the boundary results in a very rapid movement of the boundary so that relocking is avoided and the boundary velocity is almost independent of the impurity concentrations, closely resembling that observed in pure material where v = MG, M being the clean boundary mobility.

At the higher concentration, C_3, the curve is quite complex and the central part, between points b and c is unstable. Discontinuous velocity changes can be expected between the low velocities predicted by the curve between points a and b, and the high velocities predicted by the curve between points c and d. The transition would be expected to occur at a driving force between the lines ac and bd.

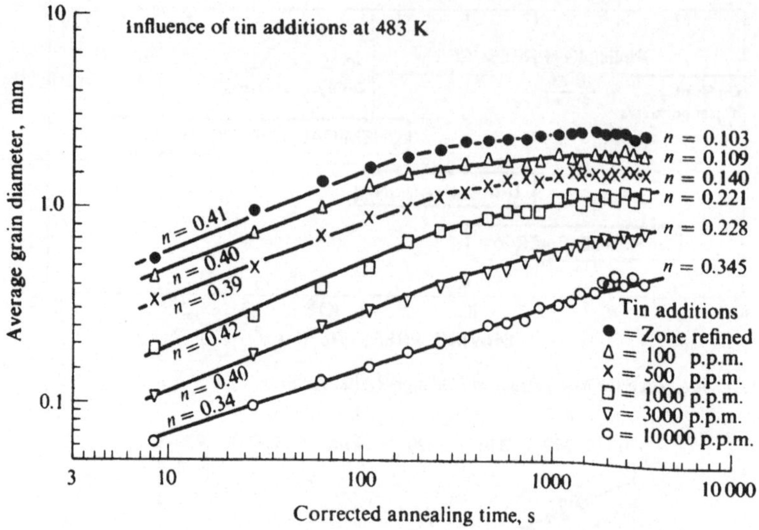

Fig. 3.12 Effect of Tin on Grain Growth in Lead. (After Doherty and Martin).

Now this theory was developed for the migration of the boundary of a new grain, during re-crystallisation, into a deformed matrix, but the mechanism should be equally applicable to the grain growth process itself. The major differences between the effects of solutes on grain growth and on re-crystallisation arise from the difference in the driving forces of the two processes. The major driving forces for microstructural change have been considered (Ashby, 1980). In general, the driving force for grain growth is two orders of magnitude less than that for re-crystallisation (which in turn can be two or three orders of magnitude less than the chemical driving forces normally associated with phase changes), Fig.3.10. Given that boundary mobilities during re-crystallisation can be marginal with respect to the release of boundaries from solutes, as shown in Fig.3.11 for a 30o (111) tilt boundary in an Fe-Au alloy (Grunwald and Haessner, 1970), then it is quite likely that in grain growth, the movement of a grain boundary will only take place if the atmosphere of solutes are carried with it. The velocity of the boundary will therefore be controlled by the impurity content as indicated by the lower driving force region of Fig. 3.9.

There is considerable experimental evidence that the rate of grain boundary migration during grain growth in lead is reduced as the concentration of tin increases, (Aust and Rutter, 1959) although special boundaries with a high degree of coincidence are hardly affected. Similar effects of tin in lead were found by Drolet and Galibois (1971), and are shown in Fig. 3.12. This effect has been observed in many other systems, but it must be remembered, that the binding energy between impurity atoms and grain boundaries will only lead to high grain boundary concentrations at low temperatures as was shown in Fig. 1.17. Heating to higher temperatures will reduce the grain boundary concentrations and therefore the effects of solute drag.

REFERENCES

Aboav D., 1970. Metallography, vol.3, p.383.

Anderson M. P., Grest G. S., and Srolovitz D. J., 1989. Phil Mag. (B) vol.59, p.293. Anderson M. P., Srolovitz D. J., Grest G. S. and Sahni P. S., 1984. Acta. Met, vol. 32, p.783.

Ashby M. F., 1980. Proc. 1st R.I.S.O., International Symposium on Metals and Materials Science, Eds. H. Hanson, A. R. Jones and T. Leffers, Roskilde, Denkark. p. 325.

Aust K. and Rutter J. W., 1959 (a) Trans. Met. Soc. AIME vol. 215, p.119.

Aust K. and Rutter J. W., 1959 (b) Trans. Met. Soc. AIME vol. 215, p.820.

Burke J. E. and Turnbull D., 1952. Progress in Metal Physics III, Pergamon Press, London. p.220.

Cahn J. W., 1962. Acta Met. vol. 10, p.789.

Drolet J. P. and Galibois A., 1971. Met Trans, vol.2, p.53.

Ewing J. A. and Rosenhain W., 1900. Phil Trans Roy Soc. (A) vol. 193A, p.353.

Fradkof V., 1992. Materials Science Forum, vol. 94-96, p.269.

Fulman R., 1952. Metal Interfaces, ASM. p.179.

Gladman T., 1966. Proc. Roy. Soc. (A). vol.294, p.298.

Gladman T., 1980. Proc. 1st. RISO International Symposium on Metals and Materials Science, Eds. N. Hanson, A. R. Jones and T. Leffers, Roskilde, Denmark, p.183.

Gladman T., 1997. The Physical Metallurgy of Micro-alloyed Steels, Inst. of Materials, London.

Glicksman M.E, Rajan K., Nordberg J., Palmer M., Marsh S. P., and Pande C. S., 1992. Materials Science Forum, vol.94-96, p.903.

Grunwald W. and Haessner F., 1970. Acta Met. vol.18, p.217.

Heckelman I., Lucke K., and Abbruzzese V., 1992. ibid., p.259.

Hillert M., 1965. Acta Met, vol. 13, p.227.

Hull F. C. and Houk W. J., 1953. Trans AIME., vol.197, p.565.

Hunderi O. and Ryum N., 1992. Materials Science Forum, vol.94-96, p.89.

Ling S., Anderson M. P., Grest G. S. and Glazier J. A., 1992. ibid. p.39.

Lucke K. and Detert K., 1957. Acta Met, vol.5, p.628.

Lucke K. and Stuwe H.P., 1963. Recovery and Re-crystallization of Metals, Ed. L. Himmel, Interscience Publishers, New York, p.171.

Matsuura K. and Itoh Y., 1992. Materials Science Forum, vol. 94-96, p.331.

Nagai T., Ohta S. and Kawasaki K., 1992. ibid. p. 313.

Nakashima K., Nagai T. and Kawasaki K., 1989. Journ Stat. Phys. vol. 57, p.579.

Pande C. S., 1987. Acta Met., vol 35, p.2671.

Pande C. S., 1992. Materials Science Forum, vol.94-96, p.331.

Rade L. and Westgren B., 1992. Mathematics Handbook, Studentliterattur, Lund, Sweden.

Smith C. S., 1952. Metal Interfaces, ASM, Cleveland, p.65.

Thorvaldsen A., 1992. Materials Science Forum, vol.94-96, p.361.

Von Neumann J., 1952. Metal Interfaces, ASM, Cleveland, p.75.

Weaire D., 1974. Metallography, vol.7, p.157.

Zener C. See S. C. Smith, 1948. Trans Met Soc. AIME, vol.175, p.345.

CHAPTER 4

Inhibition and Control of Grain Growth

4.1 INTRODUCTION

Although grain growth is a commonly observed phenomenon in crystalline materials, there were instances when grain growth ceased for no obvious reason. This led to considerable speculation about a 'limiting grain size' in relation to the overall dimensions of the polycrystalline body. Such speculation has now been discarded now that a better understanding of the grain growth process has been obtained. On the other hand, the production of fine grained steels, using small aluminium additions, was common practice in the early 20th Century, (less than 50 years after Bessemer's revolutionary introduction of bulk steel-making) even though the mechanism by which the aluminium refined the grain size was to remain a mystery for a further 50 years. It was left to Clarence Zener, a gifted physicist, to make the earliest description of the effect of second phase particles on the inhibition of grain growth (Zener, 1949). His contribution, and the subsequent developments, are described in this chapter.

The previous chapters have illustrated the nature of grains and grain boundaries, and the importance of grain boundary energy as a driving force for grain growth. However, as indicated above, the observation of limiting grain sizes in relation to the size of the polycrystalline body pointed to growth inhibiting effects other than second phase particles. Many of these observations were made on thin strip materials when the grain size was of a similar order of magnitude to that of the strip thickness. It is salutary to realise that the free surface area of the grains exceeds the internal grain boundary area when the mean linear intercept grain size is approximately equal to the strip thickness. When the grain size is near to, or larger than this value, then due account has to be given to the variation in free surface energy with grain orientation relative to the strip surfaces. This in itself would not necessarily prevent further grain growth, but when the driving force for grain growth is small (as at very large grain sizes), and the boundaries are slow moving, then thermal grooving occurs in an attempt to develop equilibrium angles between the grain boundary and the free surfaces of the two grains on either side of it. The grain boundary then has to increase its length by an amount equal to the depth of the groove if it has to move away, thereby providing a hindrance to the movement.

A third process of grain growth inhibition is seen in two phase microstructures when the second phase is present in substantial quantities and is constitutionally stable. Although this description could fit particulate second phase particles, in the so-called two-phase structure, the grains of both phases usually have similar sizes. This subject will be considered only briefly.

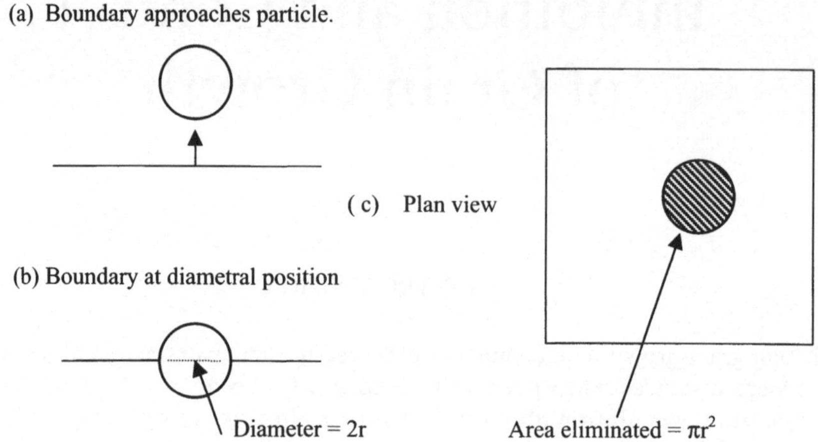

Fig. 4.1. Interaction Between Particle and Grain Boundary.

4.2 SECOND PHASE PARTICLES

The essential interaction between a second phase particle and a grain boundary arises from the disappearance of a section of the grain boundary where it intersects the second phase particle, as shown in Fig. 4.1. (a) shows a spherical particle and a planar grain boundary in elevation as separate features, with the boundary moving towards the particle. (b) shows the boundary between two grains in a position where it meets the particle diametrically. The missing element of grain boundary is indicated by the dashed line. (c) Shows the plan view where the missing area of boundary is obviously given by πr^2. Movement of the boundary away from this position requires an additional amount of energy to create this area of grain boundary, thus giving a drag effect. The boundary distorts in an attempt to avoid the creation of this boundary area and its associated energy, as shown in Fig. 4.2. This phenomenon has been described as Zener pinning in the metallurgical literature. By assuming that the boundary distortion varies inversely with the distance from the particle, (as the profile in Fig. 4.2 suggests) and that the grain boundary is normal to the particle surface at the line of contact, it has been shown (Gladman, 1966), that the change in grain boundary area (and hence energy), as the plane of the boundary is displaced from the diametral position gives rise of a maximum pinning force of $\pi r\gamma$ where r is the particle radius and γ is the grain boundary energy. This is a well established relationship in the physics of surface energy. Direct evidence of the Zener drag effect is shown in Fig. 4.3. (Gladman, McIvor, and Pickering, 1971).

Considerable complications can arise if many particles are intersected by a moving boundary, in that some particles will be intersected across their diameter, some will be exerting various drag forces, whilst some particles that are just being touched by the boundary may actually help it to move forward. However, when the particles are small and relatively well spaced in the plane of the boundary, there seems to be no disadvantage in assuming that all particles lying within a distance of +r or -r from the boundary plane are exerting their maximum drag force. This may be due to the flexibility of grain boundaries,

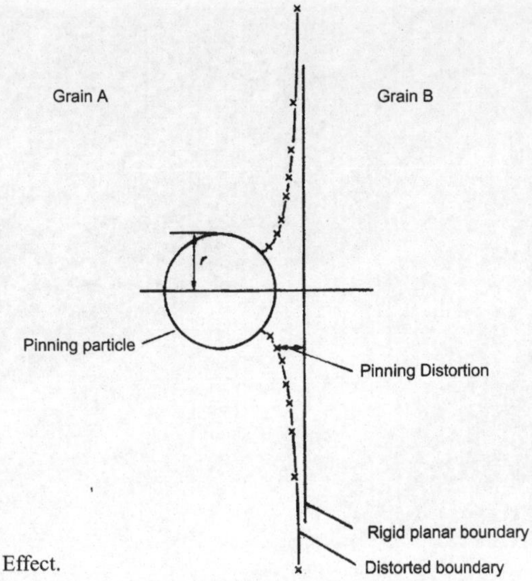

Fig. 4.2 The Zener Drag Effect.

i.e. they do not have to remain planar. Separation between particle and boundary occurs when the boundary plane moves a distance of 2r from the particle centre (Gladman, 1966). These conditions are particularly relevant when the volume fractions of second phase particles are small. In view of the significance of the Zener model, and the fact that it has never been formally published, the derivation of the Zener equation is given here.

4.1.1 THE ZENER RELATIONSHIP.

Zener was well acquainted with the physics of surface and surface energy. He started from the fact that a single spherical particle will exert a drag, P_1, on a boundary of $\pi r\gamma$, where r is the particle radius and γ is the grain boundary energy, i.e. –

$$P_1 = \pi r\gamma \qquad \text{(equ. 4.1)}$$

Given a random distribution of particles, particles will be in contact with the boundary if their centres are within + or -r of the grain boundary. The number of particles in contact with unit area of the grain boundary, N_A, will be those particles whose centres lie in a volume, V, given by unit area of the boundary and having a thickness of 2r (r on each side).

$$N_A = VN_V = 2rN_V \qquad \text{(equ. 4.2)}$$

where N_V is the number of particles per unit volume

and V is the volume defined by 2r x unit area.

The number of particles per unit volume can be derived from the volume fraction, f_v, and the particle radius, r, from –

$$f_v = N_v \,.\, 4\pi r^3 / 3 \qquad \text{(equ. 4.3)}$$

Substituting for N_v in equation 4.2 gives –

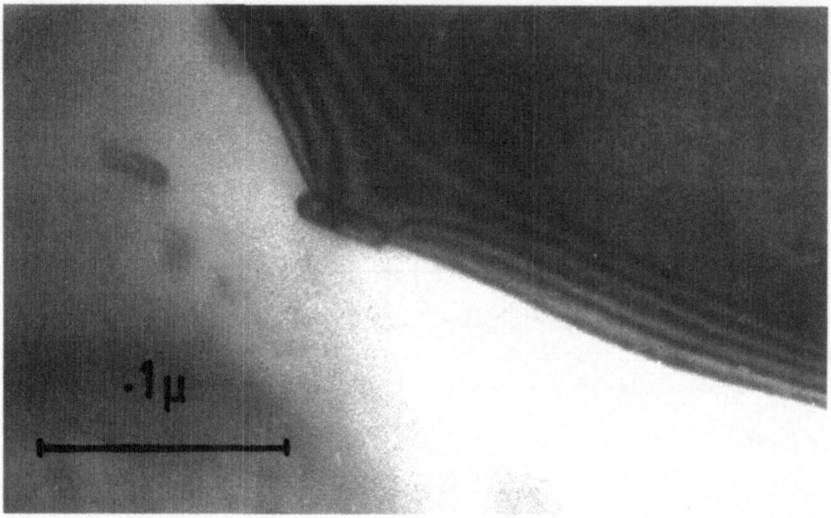

Fig. 4.3 Zener Drag. Thin foil electron micrograph showing a vanadium nitride particle dragging on a ferrite – ferrite grain boundary in a cold worked and recrystallised steel heated to 600°C.

$$N_A = 6rf_v / 4\pi r^3 \qquad \text{(equ. 4.4)}$$

If each particle exerts maximum pull on the boundary, the total drag force, P, per unit area of boundary is given by –

$$P = N_A . P_1 = 6\pi r^2 \gamma f_V / 4\pi r^3 = 3\gamma f_V / 2r \qquad \text{(equ. 4.5)}$$

In looking for a driving force for boundary movement, Zener intuitively (or from his physics training) considered a sphere, which differs from a bubble in only having one interface rather than the two (inside and outside) of the bubble. The area of this surface, A, is given by –

$$A = 4\pi R^2 \qquad \text{(equ. 4.6)}$$

where R is the radius of the sphere,

and the rate of change of surface energy with R (accompanying movement) is

$$dE/dR = (dA / dR)\gamma = 8\pi R\gamma \qquad \text{(equ. 4.7)}$$

Expressing this as the force per unit area, F, gives –

$$F = \gamma(dA/dR) / 4\pi R^2 = 2\gamma / R \qquad \text{(equ. 4.8)}$$

Now the pinning force, P, will oppose normal movement of the boundary (expansion or contraction of the sphere) so that if the force for pinning, P, exceeds that for boundary movement, the boundary will be pinned (immobile), whereas if the pinning force is less than that for movement, the boundary will be mobile. The condition of equality between the pinning force and the driving force thus provides the borderline between pinning and boundary movement. Equations 4.5 and 4.8 give this equality as –

$$3\gamma f_V / 2r = 2\gamma / R$$

$$\text{or } R = 4r / 3f_V \qquad \text{(equ. 4.9)}$$

This is the Zener relationship. This was a highly significant finding, and metallurgists were happy to see a relationship that showed that boundaries could be pinned at smaller grain sizes by increasing the volume fraction of particles of a given size, or decreasing the size of particles at a given volume fraction. Zener was happy because he had provided an insight for the metallurgists, using basic physical principles and an order of magnitude estimate of the driving force for grain growth. This equation gave the necessary understanding of grain refinement for the metallurgists to develop grain-refined materials, and remained unchanged for about 20 years.

However, it is relevant to examine the conditions involved in the Zener relationship. The particles are spherical, uniformly sized, and randomly distributed. The area, and therefore the surface energy of an isolated spherical grain actually increases with increasing grain radius (Equ. 4.7), so that the Zener equation describes the pinning required to stop an isolated spherical grain from shrinking. Further, the force for shrinking increases as the spherical grains become smaller (Equ. 4.8), so that more or finer particles are required to pin smaller spherical grains.

4.2.2 Granular Arrays with Small Volume Fractions of Uniformly Sized Randomly Distributed, Second Phase Particles

The driving forces for the growth of a tetrakaidecahedral grain of radius R in a matrix of tetrakaidecahedra of radius R_o and for the collapse of a tetrahedral grain in a radius R_i to form a 4 rayed junction are shown in Equations 3.21 and 3.27 respectively. Using these energy changes and coupling them with the Zener pinning effect allows the particle fraction and particle size requirements to be specified to inhibit grain boundary movement for these grain geometries.

(A) Growing Tetrakaidecahedra

Adding the rate of reduction in energy per unit area of the surface of the tetrakaidecahedron, Equation 3.17, to the maximum pinning force, Equation 4.5, will give zero at the borderline between inhibition and growth (i.e. the driving force for grain growth is equal to the pinning force), so that –

$$(2 / R - 3 / 2R_o)\,\gamma + 3\gamma f_V / 2r = 0 \qquad \text{(equ. 4.10)}$$

Equation 4.10 can be simplified to describe the matrix grain size that can be stabilised by a volume fraction, f_V, of particles of radius, r, in the presence of larger grains of differing size advantage, bearing in mind that if R is $< 4R_o / 3$, none of the large grains will initiate the growth process. If we denote the size advantage of the larger grain by Z $(= R/R_o)$, then Equation 4.10 becomes –

$$R_o = (1 - 4 / 3Z)\, r / f_V \qquad \text{(equ. 4.11(a))}$$

It is clear that Z values less than 4/3 will give a negative grain size, i.e. a negative R_o. If the large grains have a size advantage over their neighbours of more than 4/3, there will be a real solution for R_o.

Alternative expressions can be obtained for the minimum volume fraction of particles required to inhibit growth given the matrix grain size and particle size, or for the maximum

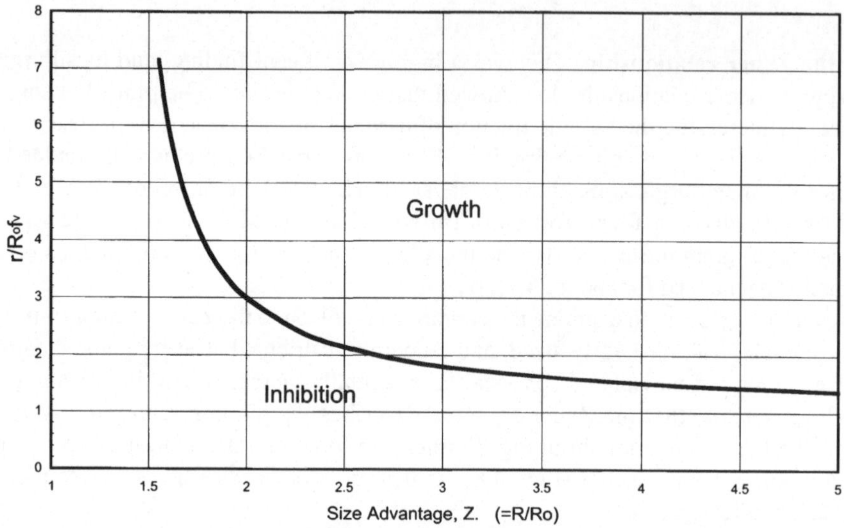

Fig 4.4 The Effect of Size Advantage on Grain Growth Inhibition.

size of particles given the matrix grain size and volume fraction of particle, viz

$$f_V = (1 - 4/3Z)\, r / R_o \qquad \text{(equ. 4.11(b))}$$

$$\text{or } r = R_o f_V / (1 - 4/3Z) \qquad \text{(equ. 4.11(c))}$$

Whilst Equation 4.11 affords a description of the growth of a large grain in a matrix of smaller grains, the incorporation of the size advantage of the larger grain, expressed as $Z = R/R_o$, permits an examination of the effects of the size advantage. The effect of size advantage on the relationship between volume fraction, particle size, and matrix grain size is shown in Fig. 4.4, where the critical conditions for the onset of grain growth are shown as a function of the size advantage. The value on the ordinate, q, is the value of $r/R_o f_V$ for different values of the size advantage $Z = R/R_o$, so that the relationship between these variables at the borderline between coarsening and inhibition is given by –

$$r = q_Z R_o f_V \qquad \text{(equ. 4.12)}$$

There are a number of interesting features shown in Fig. 4.4. Very small volume fractions of particles are required to inhibit grain growth when Z values are low (i.e. approach 4/3). Alternatively, from Equation 4.11(a) to (c) one could say that very fine matrix grain sizes could be stabilised or that the particle size could be relatively coarse when Z is small. The important feature of Fig. 4.4 however, is that the relationship is very sensitive to changes in the size advantage. If Z ($= R/R_0$) is increased from 1.40 to 2, the value of q_Z is changed by a factor of about ten. This in turn implies that almost 10 times the volume fraction of particles is required to maintain a given matrix grain size if Z is changed from 1.4 to 2.0. It is also clear from Fig. 4.4 that the sensitivity of the relationship to the actual value of Z decreases as Z values are increased. For example, if the size advantage is increased from Z = 3 to Z = infinity, the value of q is changed by less than a factor of two. Because of the changing sensitivity to Z, and the important implications for

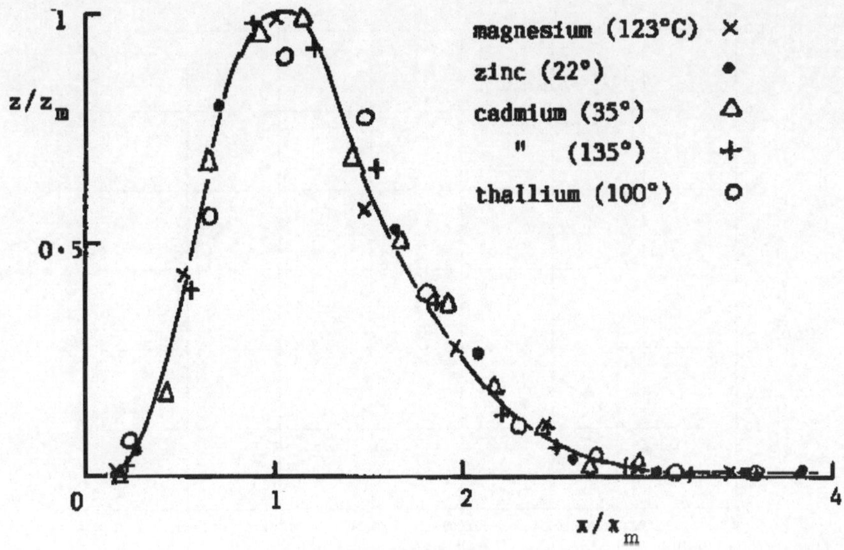

Fig. 4.5 Distribution of Grain Radii in Planar Sections of Different Metals.

the location of the borderline between grain growth and inhibition at the lower values of Z, it is pertinent to estimate realistic values of Z.

There have been many estimates of the grain size distribution of polycrystalline aggregates, most of which are based on planar intercept distributions. A particularly thorough examination of planar intercept distributions, (Aboav, 1992) for a range of polycrystalline metals, has shown that the largest grains present in significant numbers are less than three times larger than the average grain diameter, Fig. 4.5. This in itself does not necessarily give a precise size advantage as this would depend on the specific arrangement of large and small grains. From the Weaire – Aboav Law, it might be expected that large grains, would be surrounded by smaller grains and that the size advantage might be greater than three. However, planar sectioning will show a much reduced apparent grain diameter, by a factor of almost two (Table 2.4), so that the size advantage may be significantly lower at 1.5.- 2.0.

In earlier work (Gladman, 1966), values of the operative size advantage were obtained using Equation 4.11. This equation does not describe the relationship between particle distributions and grain size. Particles may be more numerous than are required to stabilise the grain size and the grain structure will be stable. Alternatively the particles may be inadequate and the grains may be growing. Equation 4.11 describes the borderline between the 'grain growth' and 'inhibition' domains. By heating samples to different temperatures to reduce the volume fraction and increase the particle size it is possible to find a temperature (and exposure time) at which grain growth starts. This should coincide with the borderline conditions implicit in Equation 4.11. If the particle sizes and frequencies and the grain size are measured after exposure at the temperature at which grain growth is about to start or has just started, substitution of the values into Equation 4.11 should yield values of Z. The results for several steels with grain refining additions of niobium carbide or aluminium nitride are portrayed in Fig. 4.6. Because the grain sizes prior to the onset of grain growth were all very similar and in the range ASTM 7/8, a constant grain radius of 14μm was inserted in Equation 4.11, (i.e. a nominal diameter of 28μm – see Table 2.8). Although Gladman (1992) noted that

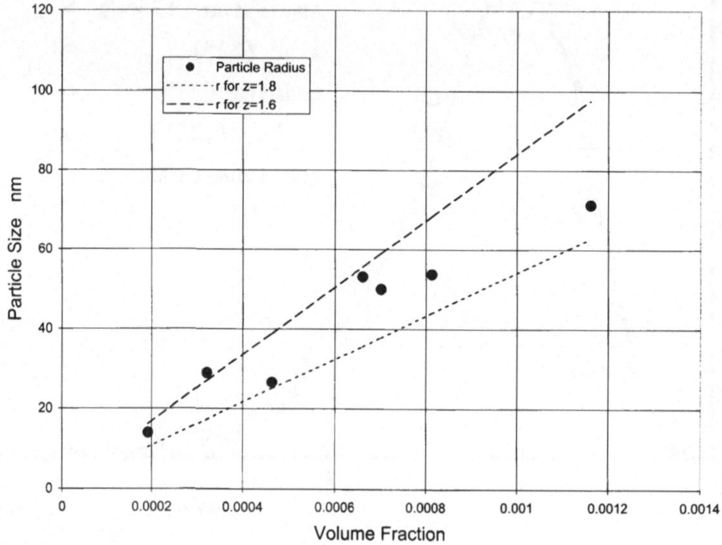

Fig. 4.6 Critical Particle Size as a Function of Volume Fraction.

the results were consistent with Z values between √ 2 and 2, it can be seen from Fig. 4.6 that the calculated values lie in an even narrower range between about 1.6 and 1.8.

It should be emphasised that the parameter Z is not a convenience factor, but represents a real variation in grain size in the polycrystalline aggregate, specifically, the ratio of the radius of the subject grain to the radius of its neighbours. Whilst it is accepted that the radii of the neighbouring grains may vary somewhat, the important grains in the furtherance of grain growth are the larger ones, and these are most likely to have 'average' sized neighbours according to the distributions shown in Fig. 3.5. There is, therefore, little point in trying to obtain a precise value of Z as it is likely to vary from specimen to specimen. Calculations of grain refining additions required to stabilise a given fine-grained structure should be made on the basis of probable/possible variations in Z. High values, e.g. Z = 2 should be used if grain growth must be avoided. On the other hand, if grain growth is desirable, then a minimum value of Z – 1.4 could be used to define maximum levels of second phase particles that could be tolerated without inhibiting grain growth.

To summarise, large grains will grow into their neighbours when the volume fraction of particles is less than that calculated from the particle size and the matrix grain size using Equation 4.11. The critical volume fraction increases as the size advantage of the large grain (relative to its neighbours) is increased. For a given volume fraction of particles, there is a critical particle size above which grains of a certain size advantage will grow, this decreasing with increasing size advantage or decreasing matrix grain size.

(B) SHRINKING TETRAHEDRA

Equation 3.27 describes the rate of lowering of energy per unit decrease in the inradius (radius of a inscribed sphere) of a shrinking tetrahedron, and is expressed in terms of unit area of the shrinking surfaces. In order to prevent grain shrinkage, sufficient particles should be added to provide a Zener drag per unit area of the surface to inhibit grain

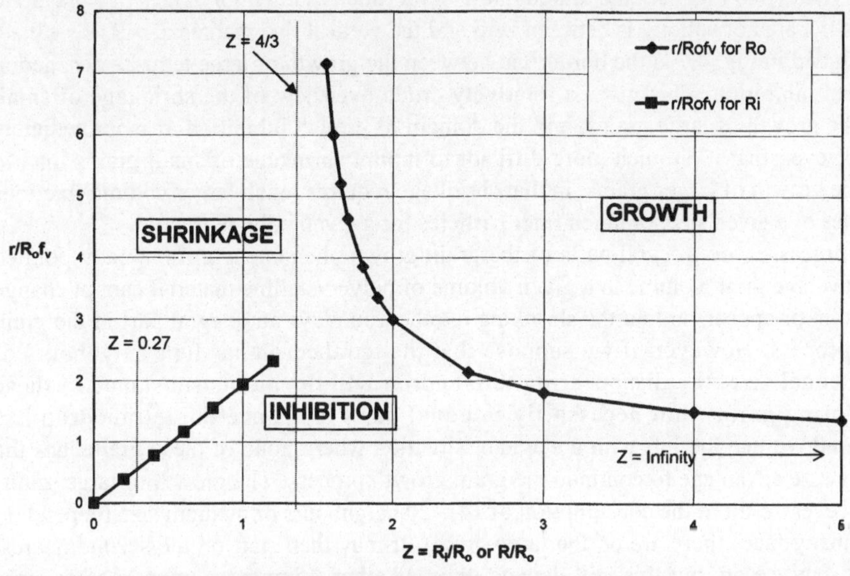

Fig. 4.7 Domains for Growth, Shrinkage, and Inhibition.

boundary movement. In other words the drag force given by Equation 4.5 must be equal to, or greater than the driving force given by Equation 3.27, i.e.

$$P = \text{or} > dE/dR_i \qquad \text{(equ. 4.13)}$$

If P is less than the driving force then grain growth will occur. The boundary between growth and inhibition, from Equation 3.27 and 4.5 is given by –

$$3\gamma f_V / 2r = (2 - \sqrt{(3/2)})\, \gamma/R_i \qquad \text{(equ. 4.14)}$$

where f_V is the volume fraction of particles,
r is the (spherical) particle radius,
γ is the grain boundary energy,
and R_i is the inradius of the tetrahedron (radius of the inscribed sphere).

This can be re-arranged to give –

$$r / f_V = R_i / (4/3 - \sqrt{(2/3)}) \qquad \text{(equ. 4.15)}$$

The significant features of this equation are the linear relationship between r/f_V and R_i with a proportionality constant of 1.935 (approximately 2), and the relationship, is entirely independent of the matrix grain size, the particulate dispersion being entirely related to the tetrahedron. However, as pointed out in Chapter 3, Section 3.2.1, the maximum inradius of a tetrahedron centred at a corner between four tetrakaidecahedra will be about 0.27 R_0. The border between shrinkage and inhibition, although independent of R_0 can be shown in terms of R_i/R_0 in order to reflect the limit of the size of the tetrahedron, simply by dividing both side of Equation 4.15 by R_0, i.e. –

$$r / R_0 f_V = 1.935 \; R_i / R_0 \qquad \text{(equ. 4.16)}$$

This is shown in Fig. 4.7, and although the borderline is shown for values of R_i/R_0 up to 1.2, the strict area of validity is between zero and the vertical line marked $Z = R_i/R_0 = 0.27$. Also included in Fig. 4.7 is the borderline between the growth of large tetra-kaidecahedral grains and inhibition. This gives a relatively crude overview of the shrinkage of small grains, the growth of large grains, and the domain of growth inhibition. It is immediately clear, however, that it is much more difficult to inhibit shrinkage of small grains than to inhibit the growth of larger grains, in that shrinkage requires much larger volume fractions of particles of a given size, or much finer particles for a given volume fraction.

Both processes are integral parts of the grain grown phenomenon. As already pointed out, the average grain volume in a given volume of polycrystalline material cannot change unless grain disappear, so that the shrinking tetrahedron plays an integral part in the grain growth process. However, if we suppose that the tetrahedron has four neighbours of roughly equal size, the disappearing tetrahedron will donate its substance to these neighbouring grains, not necessarily in equal parts, but once the tetrahedron has disappeared we may be left with a stagnant situation where none of these grains has the necessary size advantage to continue the grain growth process. The growth of large grains may, however, result in the consumption of 14 – 20 neighbours of medium size (depending on how many faces there are on the large grain). It may then start on the second nearest neighbour and so on, but this will depend upon whether it impinges upon another large growing grain, when a different citerion may come into play, i.e. does the growing grain still have the necessary size advantage over its new neighbours. Even if it does, the matrix of grains into which it is growing will become coarser and so the growth process will slow down as the driving force is reduced.

4.2.3 Granular Arrays with Small Volume Fractions of Uniformly Sized Grain Boundary Particles

In Section 4.2.2 above, the distribution of second phase particles was assumed to be random. Such a distribution might be expected if the second phase particles are precipitated prior to the formation of the grain structure, whether the granular array is then developed by cold work and re-crystallisation, by hot working, or by phase transformation. Under these circumstances the limiting matrix grain size is inversely related to the volume fraction of second phase particles, all other things being equal. Interestingly, computer modelling of grain growth in a 2-D system, (Srolovitz, Anderson, Grest, and Sahni, 1984) showed that the limiting grain size was related to the reciprocal of the square root of the volume fraction of second phase particles. Inspection of the micro-structural evolution showed that most of the second phase particles were associated with the 'grain' boundaries of the Potts model. Although there was some uncertainty about the cause of the inverse root relationship, it was subsequently shown, (Doherty, Srolovitz, Rollett, and Anderson, 1987) that it resulted from the grain boundary location of the second phase particles, and not from the two dimensional model being used. The effects of grain boundary precipitation, following the methods of Section 4.2.2 (Gladman, 1992) may be described as follows.

For the growing tetrakaidecahedron and the shrinking tetrahedron, the rates of energy change are as described by Equations 3.21 and 3.27 respectively. However, the pinning force exerted by a volume fraction, f_V, of spherical particles of uniform radius, r, will be considerably greater than when the particles are randomly distributed and only a small fraction are in contact with the boundary. Each particle exerts the same Zener drag.

Table 4.1 Effects of Grain Boundary and Random Particles on the Stabilised Grain Size.

Vol. Fraction f_v	Grain Radius, R_o, µm	
	Random Particles	Grain Boundary Particles
0.0001	166.7	5.0
0.0002	83.3	3.5
0.0004	41.7	2.5
0.0010	16.7	1.5
0.0020	8.3	1.1

Heterogeneity Z = 2; Particle Size r = 50nm

$$P_1 = \pi r \gamma$$

The number of particles per unit area of grain boundary, N_A, can be obtained from the number of particles per unit volume, N_V, and the area of grain boundaries per unit volume, A_V. In this case all of the particles occur on grain boundaries so that –

$$N_A = N_V / A_V \qquad \text{(equ. 4.17)}$$

And the total drag is given by –

$$P_g = N_A P_1 = N_A \pi r \gamma \qquad \text{(equ. 4.18)}$$

The suffix 'g' is used to denote grain boundary particles. As noted earlier –

$$N_V = 3f_V / 4\pi r^3 \qquad \text{(equ. 4.19)}$$

$$\text{and } A_V = 3 / 2R_0 \qquad \text{(equ. 4.20)}$$

where R_0 is the radius of the tetrakaidecahedral matrix.

The use of R_0 in Equation 4.20, reflects the overall grain boundary area per unit volume of the poly-crystalline aggregate, and it is implicit in this argument that the particles are randomly distributed on the grain boundaries irrespective of the local concentration, resulting from grain size variations.

Substitution of Equations 4.17, 4,19, and 4.20 into Equation 4.18 gives –

$$P_g = (3f_V / 4\pi r^3)(2R_0 / 3)(\pi r \gamma) = R_0 f_V \gamma / 2r^2 \qquad \text{(equ. 4.21)}$$

A comparison of the pinning force per unit area of grain boundary, when the particles only occur on grain boundaries (i.e. P_g in Equation 4.21) with the pinning force when the particles are randomly distributed (P in Equation 3.19) shows that – $P_g/P = R_0/3r$, and as R_0/r is typically of the order of 10^3, then the pinning force is much greater, about 300 times greater, when the aggregate contains only grain boundary particles.

(A) Growing Tetrakaidecahedra

For grain boundary particles, the borderline between grain coarsening and inhibition due to grain boundary pinning is obtained by summing the two forces and equating to zero. In the case of a large tetrakaidecahedral grain, G, in Equation 3.21, is added to the pinning force, P_g, in Equation 4.21, to give –

Table 4.2 Effects of Grain Boundary and Random Particles On the Volume Fraction of Particles Required to Stabilise a Grain Size of R_o = 10 μm.

Particle Size r nm	Vol. Fraction f_v	
	Random Particles	Grain Boundary Particles
10	0.00033	0.000001
20	0.00067	0.000004
50	0.00167	0.000025
100	0.00333	0.000100

Heterogeneity Z = 2

$$(2/R - 3/2R_0)\gamma + R_0 f_V \gamma / 2r^2 = 0 \qquad \text{(equ. 4.22)}$$

The second term in this equation is always positive, and therefore the first term must be negative for the sum to be zero, i.e. R must be greater than $4R_0/3$. Simplifying and expressing R in terms of $Z = R/R_0$ yields –

$$R_0^2 = (3 - 4/Z)\, r^2 / f_V \qquad \text{(equ. 4.23)}$$

This shows the same functional relationship between the limiting grain size and the volume fraction of particles that was observed in the computer modelling work of Srolovitz et al, i.e. the limiting grain size is inversely related to the square root of the volume fraction of particles.

Using Equations 4.23 and 4.11, the relative efficacies of grain boundary particles and randomly distributed particle can be compared. Table 4.1 shows the grain radii that can be stabilised by different volume fractions of second phase particles, assuming that the size advantage is $Z = R/R_0 = 2$, and that the particle radius is 50nm (0.05μm). In the case of the randomly distributed particles, the grain sizes vary from coarse 83μm at $f_r = 0.0002$, to fine 17μm at $f_V = 0.001$. In the case of grain boundary particles the grain sizes are ultra-fine over the whole range of volume fractions shown. The degree of grain refinement obtained from grain boundary precipitates is some 10-30 times that obtained with randomly distributed particles over the range of particle content shown, based on the given particle size and grain size advantage. The effects of particle size and the grain boundary location of particles on the volume fractions of particles required to stabilise a fine grained structure, R_0 = 10μm, is shown in Table 4.2. This table reflects the expected dependencies on particle sizes but the volume fractions of particles for grain boundary particles are some 30-300 times smaller than those for randomly distributed particles, depending upon the particle size. The increased efficiency of particles is most pronounced at small particle sizes.

Experimental evidence for the validity of Equation 4.23 relating to grain boundary precipitated particles is sparse, but a study of grain growth in low carbon steel containing aluminium nitride precipitated on ferrite boundaries (Dasarathy and Hudd, 1967), showed reasonable agreement between observed and calculated critical particle sizes if a size advantage of Z = 2 was assumed. However, a more extensive and detailed study of ferrite grain growth in these steels would be required for a rigorous validation.

(B) Shrinking tetrahedra

The borderline between tetrahedral grain shrinkage and inhibition of boundary movement, when all of the particles occur on grain boundaries, can be obtained by equating the forces given in Equations 3.27 and 4.21.

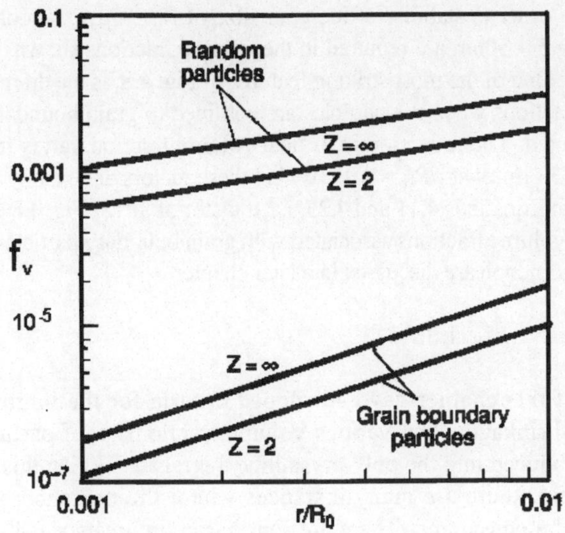

Fig. 4.8 A Comparison of Grain Boundary and Random Particles. Particle fractions can be 10 to 100 times less if they are confined to the boundaries.

$$R_0 f_V \gamma / 2r^2 = (2 - \sqrt{(3/2)})\, \gamma / R_i \qquad \text{(equ. 4.24)}$$

Equating the pinning force per unit area with the force resisting growth of the tetrahedron per unit area, is equivalent to adding the rate of change of energy due to pinning and shrinkage, and finding the shrinkage inhibition boundary when the summed energy changes are zero. Rearranging Equation 4.24 gives –

$$r^2 / f_V = R_i R_0 / 2(2 - \sqrt{(3/2)}) \qquad \text{(equ. 4.25)}$$

This equation differs from that obtained for shrinkage when the particles are randomly dispersed (Equation 4.15) because, although the driving force for shrinkage of the tetrahedron is independent of R_0, the number of particles per unit area of the boundary, (and therefore the pinning force) is directly related to R_0, see Equation 4.17. The volume fraction required to inhibit shrinkage when the particles are randomly distributed is much higher than that required when the particles are confined to grain boundaries (Equations 4.15 and 4.25) by a factor of $R_0/3r$. This is exactly the same as for the growing tetrakaidecahedron, as would be expected because both arise from the increased pinning effect when the particles are confined to the grain boundaries. As a consequence of the effects of grain boundary particles on growth and shrinkage being similarly related to the effects of randomly distributed particles, it follows that a diagram similar to Fig. 4.7 could be constructed to show the various domains of shrinkage, growth, and inhibition.

4.2.4 Comparison of Grain Boundary Particles and Randomly Distributed Particles

The limits of inhibited grain growth for randomly distributed particles and for particles confined to the grain boundaries are compared in Fig. 4.8. The relationships are expressed in Equations 4.11 and 4.23, and the critical volume fractions of particles are shown as a function of r/R_0, i.e. the ratio of the particle and grain radii, for two levels of size advantage, Z (= R/R_0). Values of r/R_0 ranging from 0.001 to 0.01 were selected as being typical of grain refined alloys (steels in particular) which require particle fractions in the range 0.0001 to 0.003, when the particles are randomly distributed. In order to stabilise a fine grain size of $R_0 = 5\mu m$, randomly distributed particles in the range r = 5 – 50nm are required in the volume fractions shown, this depending upon the size advantage. One of the most striking features of Fig. 4.8, is the dramatic reductions in the critical volume fractions when the particles are confined to grain boundaries rather than being randomly distributed. The reduction in critical volume fraction varies from about 333 times at $r/R_0 = 0.001$ to 33 times at $r/R_0 = 0.01$. Whilst these factors are exactly those given by the analytical solutions in Equations 4.11 and 4.23, i.e. a factor of $3r/R_0$, Fig. 4.8 emphasises the extremely small critical volume fractions associated with grain boundary particles. The practical applications of this phenomenon are discussed in a later chapter.

4.2.5 The Effects of Mixed Particle Sizes

The earlier sections of this chapter have developed criteria for the inhibition of grain coarsening and grain shrinkage when various volume fractions, f, of particles, having a uniform size, r, are introduced into the poly-crystalline aggregate. Whilst this facilitates the theoretical development, there are many instances where the presence of particles of different sizes need to be considered. There are four basic circumstances that can lead to the presence of differently sized particles.

(a) The precipitation process in which a given type of particle is formed will inevitably give rise to a distribution of particle sizes.

(b) If an alloy containing second phase particles is exposed at elevated temperatures (say above $0.5T_m$, where T_m is the melting point in K), Ostwald ripening can occur. In this phenomenon, small particles shrink and large particles grow, thus perpetuating differences in particle size. This phenomenon will be dealt with more fully in Chapter 5.

(c) In attempts to introduce ever increasing volume fractions of particles with limited solubility, a stage is reached where the particles begin to form at very high temperatures and become very coarse, whereas the later stages of precipitation occur at lower temperatures when fine particles are produced.

(d) Most alloys contain some impurities of limited solubility, e.g. deoxidation products, refractory materials, and so on. Each of these types of particle may exhibit their own characteristic volume fractions and size distributions. Their co-operative effects may need to be considered.

In considering the effects of particles of different sizes (Senogles, 1994; Gladman and Senogles, 1997), the Zener drag for the totality of particles has to be considered. The force opposing grain boundary motion when a single particle size is present is given by the opposing force for a single particle, P_1, in Equation 4.1, multiplied by the number of particles per unit area, N_A in Equation 4.2. If there are particles of different sizes, the overall pinning force per unit area of grain boundary is the summation of the numbers of particles per unit area for all size classes. The total pinning force, P, for particles of different sizes is given by –

$$P = (N_1 r_1 + N_2 r_2 + \text{———} N_n r_n)\, \pi\gamma \qquad \text{(equ. 4.26)}$$

where there are n particle radii, or -

$$P = \pi\gamma \Sigma N_i r_i \qquad \text{(equ. 4.27)}$$

where N_i is the number per unit area of particles of the 'i'th size group and r_i is the radius of the particles of the 'i'th size group.

Transforming the number of particles per unit area in terms of the volume fraction of particles (Equation 4.4) gives –

$$P = (3\gamma / 2)\,\Sigma\,(f_i / r_i) \qquad \text{(equ. 4.28)}$$

where f_i is the volume fraction of particles in the 'i'th size group
and r_i is the radius of the particles of the 'i'th s size group.

It should be noted that this equation relates to randomly distributed particles. If there is only one size of particle, then Equation 4.28 reverts to Equation 4.5. Balancing the pinning force, P, and the driving force for grain growth, G, (Equation 3.21) gives –

$$(3\gamma / 2)\,\Sigma\,(f_i / r_i) + (2 / Z - 3/2)\,\gamma / R_0 = 0$$

which simplifies to –

$$R_0 = (1 - 4 / 3Z) / \Sigma\,(f_i / r_i) \qquad \text{(equ. 4.29)}$$

This equation corresponds directly with Equation 4.11(a) for the single particle size, except that the term r/f is replaced by l/Σ(f/r). (Note that it is mathematically incorrect to invert the terms within the summation sign to get rid of the reciprocal).

The same procedure can be followed for particles confined to grain boundaries by expanding the total pinning force expressed in Equation 4.21 to give –

$$P_g = R_0 \gamma \Sigma\,(f_i / r_i^2) / 2$$

and coupling this with the driving force for grain growth, F, from Equation 3.21, yielding –

$$R_0 \gamma \Sigma\,(f_i / r_i^2) / 2 + (2 / Z - 3 / 2)\,\gamma / R_0 = 0$$

which simplifies to give –

$$R_0^2 = (4 - 3 / Z) / \Sigma\,(f_i / r_i^2) \qquad \text{(equ. 4.30)}$$

Again this corresponds directly with Equation 4.23 for grain boundary particles of uniform size, except that the term r^2/f is replaced by $1/\Sigma\,(f_i/r_i^2)$.

Equations 4.29 and 4.30 for the pinning criterion by randomly distributed particle and grain boundary particles respectively, relating to the simultaneous presence of various sizes of particle, show some important differences from the corresponding equations for uniformly sized particles (Equations 4.11 and 4.23). There are two main cases to consider.

CASE 1

If the particle radii show a normal distribution with a small or moderate spread of values, as shown in Table 4.3, the mean particle size and the total volume fraction can be obtained and these values used in Equation 4.11 to calculate the limiting grain size. Using the same data, it is possible to perform the more rigorous calculation where the volume fractions of particles of different sizes are used in the term $1/\Sigma(f_i/r_i)$ in Equation 4.29. As can be seen from Table 4.3, the mean particle size, r_m, is 0.02μm, the total volume fraction, f, is 0.00072, and $\Sigma(f_i/r_i)$ is $0.0321\mu m^{-1}$. Substituting the mean particle radius and the total volume fraction in Equation 4.11, over a range of size advantages of Z = 1.5 to 2.0, gives limiting grain radii in the range

Table 4.3 Calculation of the Pinning Effect from a Distribution of Particle Sizes. (*See Section 4.2.5 – CASE 1*)

r_i (μm)	n_i (μm^{-3})	n_ir_i (μm^{-2})	f_i	f_i/r_i (μm^{-1})
0.010	1	0.010	4.2x10^{-6}	4.2x10^{-4}
0.015	5	0.075	70.5x10^{-6}	47.0x10^{-4}
0.020	6	0.120	201.6x10^{-6}	100.8x10^{-4}
0.025	5	0.125	327.5x10^{-6}	131.0x10^{-4}
0.030	1	0.030	113.1x10^{-6}	37.7x10^{-4}
	Σn_i = 18	Σn_ir_i = 0.36	Σf_i 716.9x10^{-6}	$\Sigma(f_i/r_i)$ 320.7x10^{-4}

$$R_0 = 3.2 \text{ to } 9.5\mu m.$$

Using $1/\Sigma(f_i/r_i) = 0.0321\mu m$ in Equation 4.29 over the same range of size advantages ($Z = 1.5$ to 2.0), gives limiting grain radii in the range

$$R_0 = 3.5 \text{ to } 10.4\mu m$$

Although the use of the size distribution gives higher values of limiting grain size, R_0, than are obtained from assuming a constant particle size (equal to the mean particle radius), the difference is small when compared, for example, with the potential variations arising from the size advantage, Z, of the larger grains. Thus Equation 4.11 is quite adequate if the average particle size is known and not the frequency distribution of sizes.

CASE 2

Whilst the relatively narrow range of particle radii depicted in Case 1 may be relevant to many particulate dispersions, particularly where a single type of particle is found and is precipitated in a single process, there are instances where two distinctly different distributions of particle size occur. For example, a deliberate addition of niobium carbide to steel may give fine particles, which are additional to the sulphide and oxide impurities present as relatively coarse particles. Another example is often seen when titanium is added to steel. Due to the very limited solubility of titanium nitride, some precipitation of the compound occurs directly from the liquid steel, resulting in large cuboids of about 2μm edge length, whilst the remainder is subsequently precipitated in the solid steel as fine particles as small as 0.01μm in radius.

If there are two very disparate particle sizes, as shown in Table 4.4, of 0.02μm radius and 1μm radius, and there is a larger volume fraction of the very coarse particles, the situation is as follows. With the same volume fraction of particles as was used in Case 1, i.e. $f = 0.00072$, calculation of the mean particle size, r_m, gives 0.020011μm, and $\Sigma f_i/r_i$ is 0.01542μm^{-1}. Calculating the limiting grain size, R_0, below which growth would occur, from Equation 4.11 with the mean particle size and total volume fraction, and a range of size advantages of $Z = 1.5$ to $Z = 2.0$ yields –

$$R_0 = 3.1\mu m \text{ to } 9.3\mu m$$

This is almost the same result as that given for Case 1. However, using Equation 4.29, which allows for the different particle sizes through the fraction $\Sigma(f_i/r_i)$ yields –

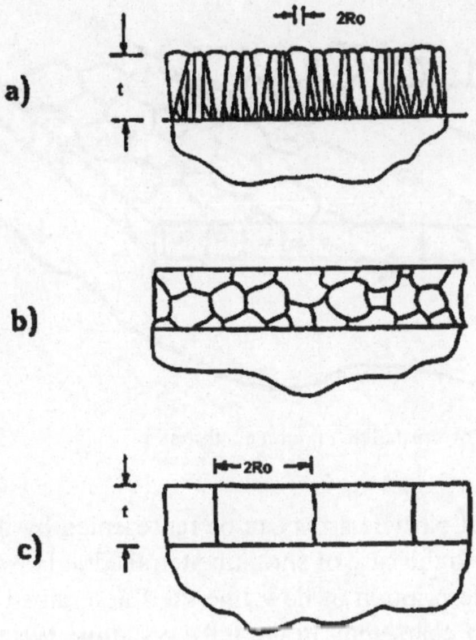

Fig. 4.9 Grain Structures in Foils. (a), As deposited. (b). Early stages of annealing. (c). Through thickness grains.

$$R_0 = 7.2 \text{ to } 21.6\mu m$$

This is a much higher result, and closely resembles that which would be found if the very coarse particles were to be ignored completely, i.e. f = 0.0003 and r = 0.02μm yielding –

$$R_0 = 7.4 \text{ to } 22.2\mu m$$

This is very different from Case 1, where the mean particle size, r_m, and the total volume fraction, f, could represent the size distributed particles. In Case 2, the coarser particles can be neglected completely.

The two cases given above represent the extremes of the frequency distributions of differently sized particles. When particles of different sizes are seen to be present, it may not always be clear whether Equation 4.11 could be taken as representative of the whole distribution using r_m and f, or whether the very coarse particles could be neglected. As will be shown in a later chapter, the real appreciation of the contribution of particles of a particular size class, i, comes from the magnitudes of f_i/r_i in relation to $\Sigma(f_i/r_i)$ as shown in Tables 4.3 and 4.4. In Case 1, (Table 4.3) particles in all class intervals contribute significantly to the sum, $\Sigma(f_i/r_i)$. In Case 2, (Table 4.4) only the small particles contribute significantly, and the coarse particles, although representing more than half the volume fraction of particles present, contribute less than 3% of the pinning action, $\Sigma(f_i/r_i)$.

Table 4.4 The Effect of Disparate Particle Sizes. (*See Section 4.2.5 CASE 2*)

r_i μm	n_i μm^{-3}	n_ir_i μm^{-2}	f_i	f_i/r_i μm^{-1}
0.02	8.9525	0.17905	0.00030	0.01500
1.00	0.0001	0.00010	0.00042	0.00042
	Σn_i = 8.9526	Σn_ir_i = 0.17915	Σf_i = 0.00072	$\Sigma f_i/r_i$ = 0.01542

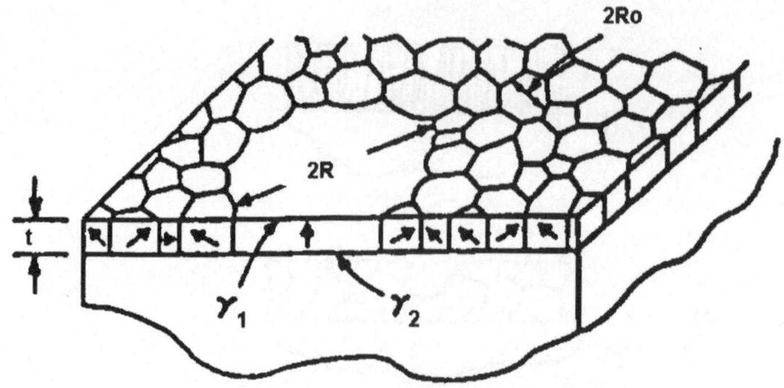

Fig. 4.10 Schematic Perspective of a Polycrystalline Film on a Substrate.

The action of distributions of particle sizes can be represented by the $\Sigma(f_i/r_i)$ function, and this would also be relevant to the case of shrinking tetrahedra. However, the shrinkage effect is more relevant to the description of how fine the fine grained structure is, rather than to the onset of major grain coarsening in the polycrystalline system, and will not be considered further here.

4.3 FREE SURFACE EFFECTS

In the past decade or two there have been several scientific and technological developments that have had important effects on the study of grain growth, and the practice of grain control. Controlled deformation and thermal cycles have been used to develop crystallographic texture in thin strip products in order to improve cold·formability or magnetic properties of ferrous materials for a considerable period of time. Developments in steel-making processes have led to significant reductions in impurity levels, with a consequent reduction in both particle inhibition of grain growth and solute drag effects on grain coarsening kinetics. These effects have resulted in vast changes in the attainable grain structures in strip products. For example, in silicon steel strip for magnetic applications, strip thicknesses are typically of the order of 0.3mm; in these same products, grain diameters of several centimetres are not uncommon.

In recent years, considerable interest has been shown in the manufacture of rapidly solidified ribbons, and also patterned thin films used as interconnects in integrated circuits. Films as thin as 10-100nm are of interest, and may be created by electro-deposition or by vapour deposition processes onto a substrate.

The important features of thin strip or films arises from the grain size being of the order of, or larger than, the strip or film thickness. In bulk materials, the grain size is orders of magnitude less than the stock size and the three dimensional network of grain boundaries is paramount in controlling the grain growth process. This situation can also prevail in the initial stages of growth in strip and film when the initial structure is extremely fine grained, as shown in Fig. 4.9 (Thompson, 1992). The deposits may initially be columnar, as shown, or even amorphous. The early stages of annealing may give a polyhedral grain structure, Fig. 4.9(b), in which bulk grain growth phenomenon will dominate. However, prolonged annealing may then produce a grain structure in which all grains occupy the full strip

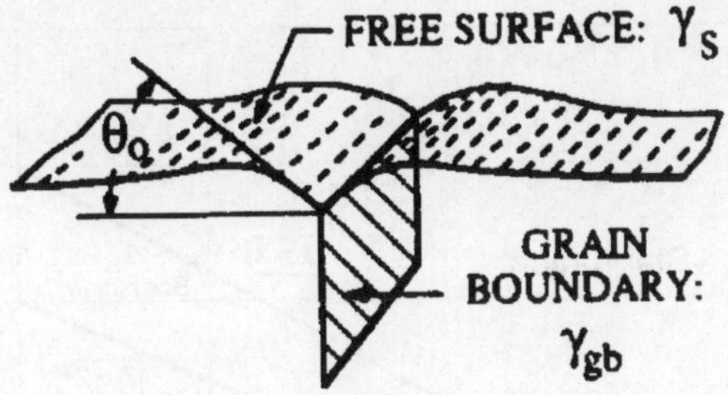

Fig. 4.11 Geometry of a Grain Boundary Groove. (Mullins, 1958).

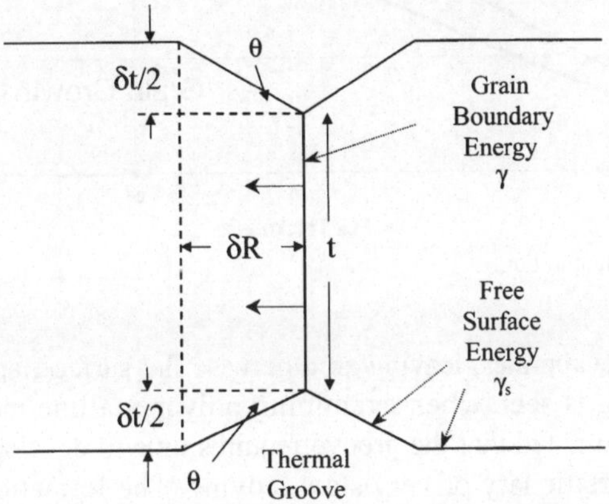

Fig.4.12 Change in Width of Foil, δt, When Grain of Radius R Grows by an Amount δR.

thickness and the grain boundaries are substantially normal to the surface, Fig. 4.9(c). Under these circumstances, the grain structure will appear to be two dimensional if viewed normal to the surface of the strip, and has been treated as such from a mathematical standpoint. However, it the strip is viewed from any other direction, the three dimensional nature of the structure is clearly seen, Fig. 4.10.

There are two important phenomena associated with these structures, arising from the free surface energy. These are –

(i) Thermal grooving arising from triple point equilibration.

(ii) Grain growth driven by differences in surface energy.

4.3.1 Thermal Grooving.

In the structures denoted by Fig. 4.9(c) where grains occupy the full strip thickness, the emergence of the grain boundary at the free surface provides a triple point as depicted in Fig. 4.11, (Mullins, 1958). Following the equilibration of forces between the grain boundary energy, γ, and the surface energy γ_s, it is relatively easy to show that –

$$\gamma = 2\gamma_s \sin\theta \qquad \text{(equ. 4.31)}$$

The profile at the thermal groove formed as a result of this equilibration depends on whether the matter from the groove diffuses along the surface, giving the humped profile

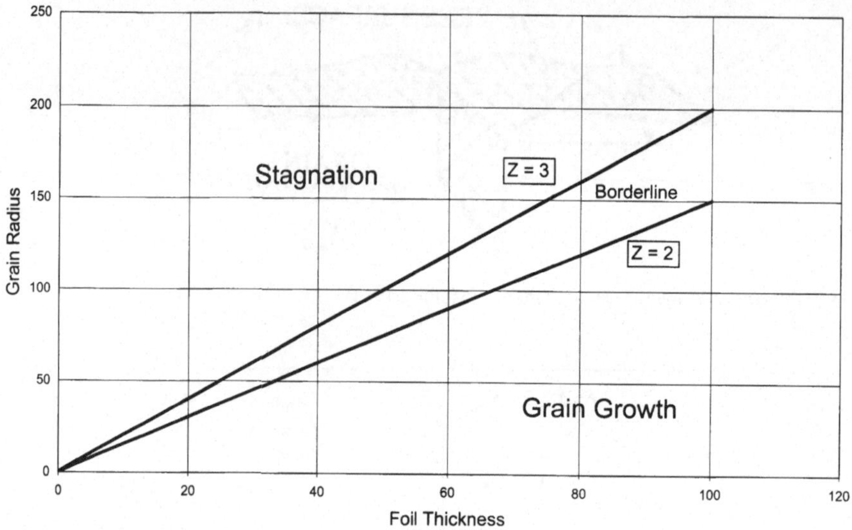

Fig.4.13 Limiting Grain Radius in Foil.

shown in Fig. 4.11, or evaporates, leaving an otherwise flat surface, apart from the groove. Such thermal grooving is seen when examining polycrystalline metals on a hot-stage microscope. It should be noted that the groove requires time to develop, which implies that the boundary must be stationary or very slow moving. The formation of such a groove, nevertheless, provides a pinning action, in that movement of the grain boundary away from the groove requires an increase in the grain boundary area due to the increase in the grain boundary length in the strip thickness direction which occurs on both surfaces of the strip. If a grain of radius, R, undergoes growth by an amount δR, there is an increase in the area of the growing grain due to the extension of the perimeter, and an increase due to the separation from the two grooves, δt, over the whole of the perimeter. There is also a decrease in area of grain boundary due to the elimination of the boundaries belonging to the neighbouring grains that are consumed. If these energy changes are expressed as changes per unit area of the growing boundary, as was the case for the bulk model in Section 4.2.2(a) it is possible to define the borderline between grain growth and the 'stagnation' due to pinning by surface grooves. The increase in area, δA_1, due to the growing perimeter is given by –

$$\delta A_1 = 4\sqrt{3}\, t\, \delta R \qquad \text{(equ. 4.32)}$$

where $4\sqrt{3}$ arises from assuming an hexagonal grain,

and t is the thickness of the strip or foil.

The increase in area, δA_2, due to detachment from the grooves is given by –

$$\delta A_2 = 4\sqrt{3} R\, \delta t \qquad \text{(equ. 4.33)}$$

The change, δt, can be related to the increase in radius, δR, as shown in Fig. 4.12 where –

$$\delta t = 2 \tan \theta . \delta R \qquad \text{(equ. 4.34)}$$

Substituting this into Equation 4.33 gives –

$$\delta A_2 = 8\sqrt{3}\, R \tan \theta . \delta R \qquad \text{(equ. 4.35)}$$

If we assume that tan θ is approximately equal to sin θ when θ is small, and substituting from Equation 4.31, we have –

$$\delta A_2 = 8\sqrt{3}R(\gamma / 2\gamma_s)\,\delta R = 4\sqrt{3}R(\gamma/\gamma_s)\,\delta R \qquad \text{(equ. 4.36)}$$

Again, if we treat the neighbouring grains as hexagonal grains of radius, R_0, and obtain an average grain boundary density (area, A, in a volume V), we obtain –

$$A / V = 2\sqrt{3}R_0 t / 2\sqrt{3}R_0^2 t \qquad \text{(equ. 4.37)}$$

$$\text{or } A/V = 1/R_0 \qquad \text{(equ. 4.38)}$$

It should be noted that the numerator in Equation 4.37 represents one half of the grain boundary area of a single grain because of the shared interface. The volume increase when the grain of radius, R, grows to a radius, R + δR, multiplied by the grain boundary density in the matrix gives the area of grain boundary consumed, δA_3, by the growing grains, i.e. –

$$\delta A_3 = -(4\sqrt{3}Rt.\delta R) / R_0 \qquad \text{(equ. 4.39)}$$

Summing the area changes given by Equations 4.32, 4.36 and 4.39, and expressing these in terms of area change per unit area of the growing boundary gives –

$$\delta A/A = [4\sqrt{3}t\,\delta R + 4\sqrt{3}R(\gamma/\gamma_s)\delta R - 4\sqrt{3}(R/R_0)t\,\delta R] / 4\sqrt{3}Rt \qquad \text{(equ. 4.40)}$$

$$\text{or } (dA/dR)(1/A) = 1/R - 1/R0 + (\gamma/\gamma s)t \qquad \text{(equ. 4.41)}$$

Thus the rate of change of grain boundary area (and therefore energy) per unit area of the growing grain is negative only when $1/R_0$ exceeds the sum of 1/R and $\gamma/\gamma_s t$. By making the substitution $Z = R/R_0$, Equation 4.41 can be written (in the limit) as –

$$(dA/dR)(1/A) = (1/Z - 1)/R_0 + (\gamma/\gamma_s)t \qquad \text{(equ. 4.42)}$$

The partition between grain growth and stagnation will occur when the rate of change of area (and therefore energy) is zero. i.e.

$$(1 - 1/Z)/R_0 = (\gamma/\gamma_s)t \qquad \text{(equ. 4.43)}$$

If the matrix grain size is less than that required for the film thickness, surface/grain boundary energy ratio, and size advantage, $Z(= R/R_0)$, then growth will occur. If the matrix grain size attains the value obtained from Equation 4.43, then grain growth will stagnate. The situation is depicted in Fig. 4.13, where the limiting value of R_0 is shown as a function of the foil thickness. The limit is obviously affected by the grain size heterogeneity (as depicted by

Table 4.5 Comparison of Free Surface with Grain Boundary Energies.

Metal	Surface Energy γ_s Jm^{-2}	Grain Boundary Energy γ Jm^{-2}	Ratio γ_s/γ
Aluminium	1.1	0.60	1.83
Gold	1.4	0.40	3.50
Copper	1.75	0.53	3.30
Iron (bcc)	2.1	0.80	2.63
Iron (fcc)	2.2	0.79	2.78
Platinum	2.1	0.78	2.69
Tungsten	2.8	1.07	2.62

Z), and by the ratio of the surface energy to the grain boundary energy. The effect of grain size heterogeneity ($Z = R/R_0$) is illustrated for Z = 2 and Z = 3, and clearly shows the movement of the borderline between grain growth and stagnation, i.e. the effect of Z on the limiting matrix grain size for any given foil thickness. It should be noted that dimensions have been omitted for foil thickness and grain radius. The numbers are representative of foil thicknesses expressed in nanometres, or strip thicknesses expressed in micrometres, as long as the grain radius takes the same dimension as the thickness. The value of the ratio of surface energy to grain boundary energy used in Fig. 4.13 was $\gamma_s/\gamma = 3$. Values for different metals were collated (Martin and Doherty, 1976) and are shown in Table 4.5. The values of γ_s/γ are generally in the range 2.6- 3.5 with the exception of aluminium, which has ratio of 1.83, so that the value of $\gamma_s/\gamma = 3$ is a reasonable assumption. If the value of grain size heterogeneity is assumed to be Z = 2, then, as shown in Fig. 4.13, the grain radius is 1.5 times the foil thickness, and therefore the limiting grain diameter would be three times the foil thickness. This is in excellent agreement with experimental observations of the stagnant grain size (Frost, Thompson, and Walton, 1992; Beck, Holtzworth, and Sperry, 1949; and Palmer, Thompson, and Smith, 1987).

The formation of surface grooves requires exposure of the triple point at a temperature for a period of time. Initially, the larger grains in a fine grained structure, would be growing rapidly, and the opportunity to form grooves on the mobile boundaries would be limited. For average sized grains, some grooving might occur at a fairly early stage. As the average grain size of the aggregate increases, however, the growth rate of all grains will slow down, and this will allow thermal grooving to occur, with the consequent retarding influence on grain growth. It is interesting to note that the grains with the largest size advantage will be the last to stop growing, and might be expected to give some form of selective growth. However, even a huge grain (of infinite radius) would cease growing if the matrix grain size attained a size of six times the foil thickness. If there was only one very large grain when the remainder become stagnant at a grain size less than six times the foil thickness, then the abnormal form of grain growth would be expected.

4.3.2 Variations in Free Surface Energy

The concept of stagnation introduced in the preceding section strictly relates to the cessation of normal grain growth. Subsequent growth in the case of foil or thin strip is observed to take place by abnormal grain growth, or secondary re-crystallisation, in which one (or very few) grain(s) will grow in an unchanged matrix. Other treatments of the grain growth process in foils were based on curvature driven growth which omitted the importance of relative grain size (heterogeneity), and consequently did not explain the progressive loss of growing grains when their size advantage became inadequate. A further factor was sought that would favour the growth of only few selected grains, and attention was turned to the effects of surface energy. Irrespective of the reason for investigating the effects of surface energy, it is clear that variations in surface energy must contribute importantly in thin strip materials or foils. A relatively simple calculation of the grain boundary area of hexagonal grains occupying the full thickness of unit area of strip showed that it is equal to the two surface areas (top and bottom faces) when the diameter (distance across flats of the hexagon) is equal to the strip thickness. Thus, in the situation where normal growth stagnates, i.e. D = 3t, the surface area of the strip dramatically exceeds the grain boundary area.

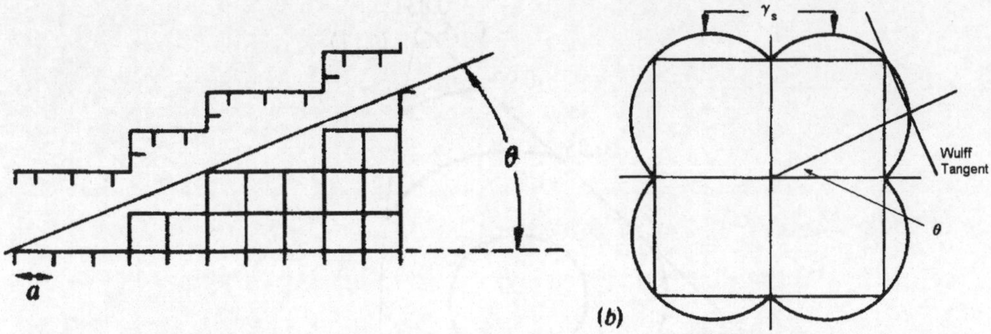

Fig. 4.14 (a). Broken Bonds on the Surface at an Angle to the (01) Plane. (b). Surface Energy Variation with Angle (the magnitude of g is measured from the origin).

The change in surface energy as an hexagonal grain of radius R grows to a radius R + δR depends upon the differences in surface energy of this grain, and those of its neighbours. If we simplify the situation so that we have a grain of radius R, with a surface energy γ_1 surrounded by grains of radius R_0 having surface energies of γ_s, the change in surface area attached to the growing grain δA_4, is given by –

$$\delta A_4 = 2\,(4\sqrt{3R})\,\delta R \qquad \text{(equ. 4.44)}$$

and the energy of this element of area changes by an amount $\gamma_1 - \gamma_s$, so that the energy change is given by –

$$\delta E_4 = 8\sqrt{3}\,R\,(\gamma_1 - \gamma_s)\,\delta R$$

When expressed in terms of energy change per unit area of growing interface -

$$\delta E_4\,(1/A) = 8\sqrt{3}\,(\gamma_1 - \gamma_s) R\,\delta R / 4\sqrt{3}\,Rt$$

In the limit, therefore –

$$(dE_4/dR)\,(1/A) = 2\,(\gamma_1 - \gamma_s)/t \qquad \text{(equ. 4.45)}$$

In order to combine this energy change with those due to thermal grooving effects and grain boundary energy effects, Equation 4.42 must be converted to energy changes and the surface energy term from Equation 4.45 added, yielding –

$$(dE/dR)\,(1/A) = (1/Z - 1)\,\gamma/R_0 + (\gamma/\gamma_s)\,\gamma/t + 2\,(\gamma_1 - \gamma_s)/t$$

$$\text{or } (dE/dR)\,(1/A) = (1/Z - 1)\,\gamma/R_0 + (2\gamma_1 - 2\gamma_s + \gamma^2/\gamma_s)/t \qquad \text{(equ. 4.46)}$$

This equation is quite complex but worthy of some examination and comment.

(i) Only when the grain has a size advantage can it cause a reduction in energy by growing, i.e. Z > 1. The value of Z can thus range between 1 and infinity, and the function (1/Z – 1) can range between 0 and – 1. Thus the first term on the RHS of Equation 4.46 has a maximum rate of change in energy of $-\gamma/R_0$, i.e. a maximum driving force of γ/R_0.

(ii) The second term on the RHS of the equation includes the drag due to thermal grooving

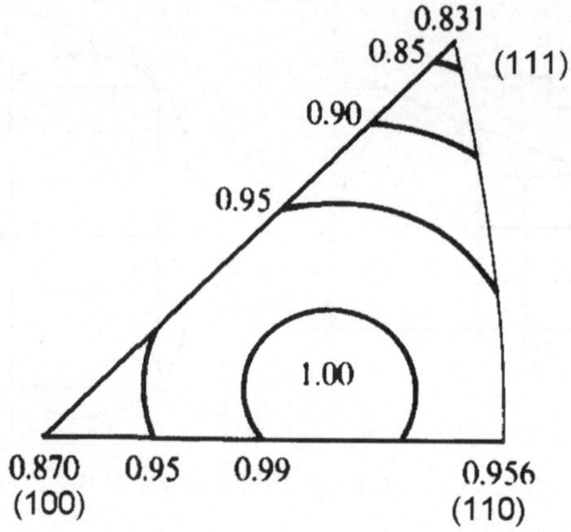

Fig. 4.15 Stereographic Projection of the Free Surface Energies for the Face Centred Cubic Crystal Structure.

represented by $\gamma^2/\gamma_s t$. Again making the assumption that $\gamma = \gamma_s/3$, from the data in Table 4.5, the groove term may be reduced to $0.11\gamma_s/t$. The significance of this drag effect was discussed in the previous section, but the overall contribution of the second term in Equation 4.46 would clearly depend upon the relative values of γ_1, γ_s, and γ.

(iii) The relative magnitudes of γ_1 and γ_s are clearly of vital importance. If γ_1 is less than γ_s, the growth of the grain of radius R will be augmented by an additional rate of energy reduction of $(\gamma_1 - \gamma_s)/t$, because $(\gamma_1 - \gamma_s)$ is negative. If γ_s is the smaller of the two however, the surface energy term will provide an additional drag as $(\gamma_1 - \gamma_s)$ will be positive. The magnitude of the difference between γ_1 and γ_s is clearly important in controlling the additional driving force or the additional drag.

The variation in surface energy from one grain to another, arising from the grain orientation with respect to the surface plane, has been described previously (Martin and Doherty, 1976), and only a brief outline will be given here. Following the calculation of Mullins, 1963, the number of broken bonds per unit area of surface, N_A, are calculated and multiplied by one half of the bond energy, U, since the break produces two surfaces. Initially this was done for a simple cubic 2-D crystal fractured at an angle θ to a close packed (01) plane, Fig. 4.14(a). The broken bonds per unit length of edge, N_L, are given by –

$$N_L = \sin\theta / a + \theta / a$$

where a, is the inter-atomic spacing.

The "surface" energy, γ_s, is then given by –

$$\gamma_s = N_L U = U(\sin\theta + \cos\theta)/2a \qquad \text{(equ. 4.47)}$$

The orientation dependence of the "surface" energy is shown in Fig. 4.14(b) in the form of a polar plot, where γ_s is the distance from the origin shown as a function of the angle θ over a full rotation of 360°. It can be seen that there are four maxima occurring at 45°, 135°, 225°, and 315°, and four cusps representing the four minima at 0°, 90°, 180°, and 270°. The ratio of maximum to minimum "surface" energy shown in Fig. 4.14(b) is thus 1.4142 (the square root of 2). This approach has been applied to three- dimensional f.c.c. and b.c.c. lattice structures, taking into account the somewhat weaker bond energies associated with second nearest neighbours, (Sundquist, 1964). The γ_s values are depicted in Fig. 4.15 in a stereographic triangle, where the full range of relative surface energies are portrayed for a f.c.c. structure, when the second nearest neighbour bond is one tenth that of the nearest neighbour bond. The calculated relative surface energies vary between 0.831 and 1.000. Some confidence can be placed on the results as the equilibrium shapes of small crystals calculated on the basis of minimum surface energy showed excellent agreement with those observed on small (1μm) f.c.c. crystals (Sundquist, 1964). Similar calculations have been made for b.c.c. structures and show similar variations in free surface energy with orientation. The basic difference is that the (111) surfaces have lowest surface energy for the f.c.c. structure, and the (110) surfaces have the lowest surface energy for b.c.c. structures. The significance of variations of the relative magnitudes shown now becomes clear. If we let the grain boundary energy term be expressed as $\gamma = \gamma_s/3$ as before, and let the ratio of the surface energy of the growing grain, γ_1, to that of its neighbours, γ_s, be Γ, then Equation 4.46 becomes -

$$(dE/dR)(1/A) = (1/Z - 1)\,\gamma_s/3R_0 + (2(\Gamma - 1) + 1/9)\gamma_s/t \qquad \text{(equ. 4.48)}$$

If the energy change is negative, those grains with the specified size advantage and favourable surface energy will grow. If the energy change is positive, then there will be no grain growth. The borderline between these situations occurs when (dE/dR)(1/A) = 0, which can be written as –

$$[6(\Gamma - 1) + 1/3]\,R_0/t = (1 - 1/Z) \qquad \text{(equ. 4.49)}$$

From Fig. 4.15, the surface energies of different planes in a f.c.c. structure, ranging from 0.831 to 1.0, would allow Γ to vary between 0.831 and 1/0.831 (i.e. 1.2030). These extremes of Γ would only occur when a grain of minimum surface energy was surrounded by grains of maximum surface energy, and vice versa. If the placements of grain orientations were random, then it is far more likely that a grain with an extremely low, or an extremely high surface energy will be surrounded by grains of an average surface energy, as these would appear to be far more frequent according to the spatial distribution shown in Fig. 4.15. Although the extreme value of Γ cannot be ruled out, values in the range from 0.831/0.92 = 0.9 to 1.0/0.92 = 1.1, would appear to be far more probable.

The relationship between borderline values of Γ and Z is shown in Fig. 4.16. This relationship is obviously affected by the value of R_0/t, and so, for the purpose of Fig. 4.16, as value of R_0/t of 1.5 has been chosen. There are several points to be made about this curve. Between points 'a' and 'b', the reduction in surface energy overcomes both the pinning by the thermal groove effect and the size disadvantage of the growing grain (i.e. Z < 1). The critical size for these 'through thickness' grains is clearly changed by having an alternative source of energy reduction, so that a smaller grain can consume its larger neighbours provided that there is a sufficient surface energy difference. At point 'b' where Z = 1, the energy reduction by having a low surface energy is that which is required to overcome the pinning effect of thermal grooves. Between points 'b' and 'c' the surface energy reduction and the size advantage, combine to overcome the thermal groove pinning, until at point 'c', where $\Gamma = 1$, the surface energy effect disappears, and the unpinning of the thermal groove is achieved from the grain size advantage effect alone, when Z = 2. Between points 'c' and 'd', where both the surface energy changes and groove unpinning oppose growth of the grain, the size advantage, with its grain boundary energy effects, has to overcome both of these. Thus, grains with Γ less than that shown by the curve, or with Z

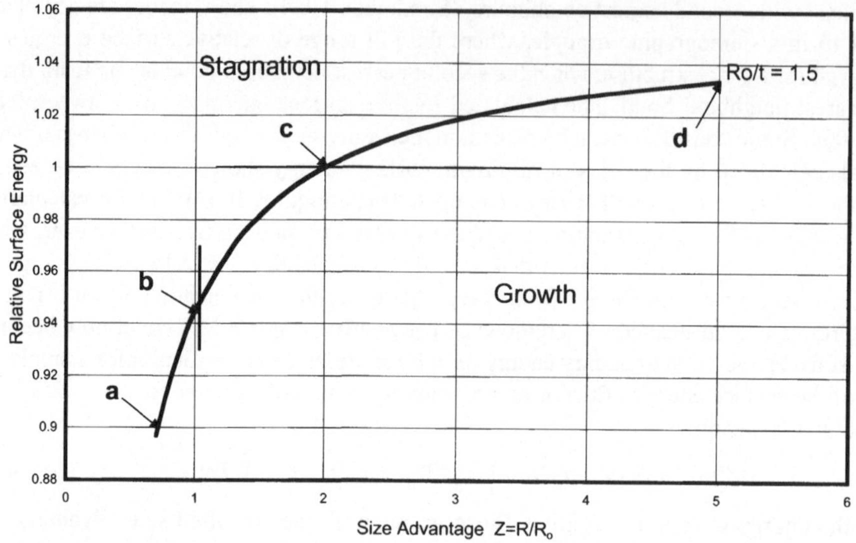

Fig. 4.16 Stagnation Domain in Thin Films.

greater than the value, will suffer growth. Otherwise the structure will be stagnant.

The zones of stagnation and growth shown in Fig. 4.16 are specific to the case where $R_0/t = 1.5$. We have already noted above, that general grain growth ceases when the matrix grain radius attains a value some 1.5 times the thickness of the foil, and that further growth occurs by an abnormal growth process. The implications of this observation is that in a polycrystalline foil, there are very few grains which lie in the growth domain of Fig. 4.16, and which are able to grow. Various reasons have been suggested for there being very few growing grains, and these include non-random distributions of grain orientations, and associated surface energies, and the presence of second phase particles (or pores). The surface energy suggestion effects are considered in the next section.

4.3.3 Substrates and the Capping of Foils

The foil structure illustrated in Fig. 4.10 also shows a substrate upon which the foil had been deposited. This, and the nature of the substrate, can dramatically modify the 'surface' energy. This surface of the foil becomes an interface having an interfacial energy which may vary from one grain to another, and which may be dramatically affected by the nature of the substrate. If the substrate is amorphous, the interface may have a single interfacial energy, γ_u, over the whole of the interface. The surface energy change accompanying grain growth shown in Equation 4.45 would then be reduced to $(\gamma_1 - \gamma_s)/t$, rather than the $2(\gamma_1 - \gamma_s)t$, when there are two free surfaces. If the upper surface is then capped with a similar material, the free surface free energy term $(2\gamma_1 - 2\gamma_2)$, would disappear from Equation 4.46 and the term $6(\Gamma - 1)$ would disappear from Equation 4.49. However, as pointed out by Thompson (1992), the interfaces between grains and the substrate in the absence of capping could have different interfacial energies which are not the same as their free corresponding surface energies, and which may not even rank in the same order as those of the free surface energies, so that certain grains could have a driving force for grain growth

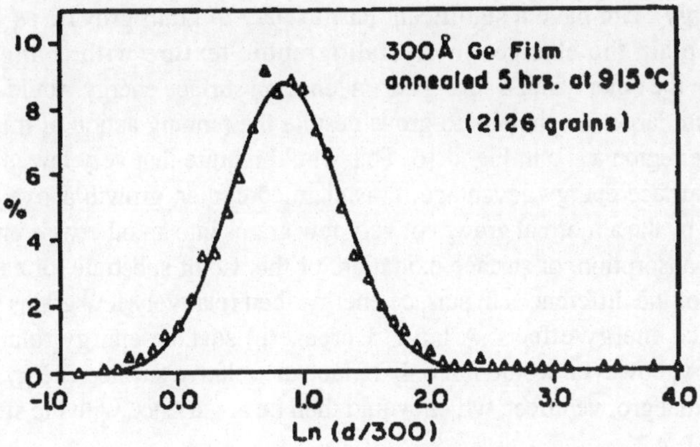

Fig. 4.17 Lognormal Distribution of Grain Diameters in a Thin (30nm) Foil of Germanium

on one side of the foil and even a drag effect on the other side.

It is worth mentioning at this stage that a single crystal substrate can produce interesting effects, in that grains in the foil with particular orientations, giving near atomic matching with the substrate, can have low interfacial energies, and may grow preferentially. This will give a textured foil where the texture is controlled by the substrate grain orientation and the lattice matching, (Thompson, 1992).

The above considerations relate to manufactured interfaces in that the substrate and capping materials are deliberately chosen and engineered. There are, however, circumstances that may arise fortuitously by the adsorption of impurity atoms on a free surface, or by oxidation of the free surface to give an extremely thin layer of oxide or other compound, depending upon the nature of the atmosphere. The effects of such changes on the surface energy, and more specifically on the surface energies of certain crystallographically oriented surfaces, have been described by Hondros and Stuart, (1968). These workers used tubular specimens of 3% Si steels, with a 25μm wall thickness, and a post anneal grain diameter in excess of 0.5mm. The grain boundary energies were determined by the zero creep rate method, and the surface energies were then derived from measurements of the thermal groove angles on wire specimens. The results were extremely complex but showed the following features –

(a) The relative surface energies of the (110) plane was 0.85 times that of the (100) plane when measurements were made at very low oxygen partial pressures. This is in excellent agreement with the calculated values quoted earlier for the b.c.c. structure (Sundquist, 1964).

(b) The surface energies associated with both the (100) and (110) planes decreased slightly with increasing temperature at low oxygen partial pressures.

(c) Increasing the oxygen partial pressure in the test chambers produced a dramatic decrease in the surface energies, (as would be expected from adsorption), reversed the temperature dependence (as would be expected from de-sorption), and inverted the relative surface energies of the (110) and (100) planes.

There are very few systems for which the detailed surface energies are known for a wide range of temperatures and atmospheres. It is clear from the above work that surface energy, (or interfacial energy), can have a significant part to play in grain growth, as it would be very difficult to explain the changes in crystallographic texture with changes in annealing atmosphere. On the other hand, a strong dependence on surface energy would even allow some small grains with larger neighbours to grow, despite the pinning action of thermal grooves, as indicated by the region a – b in Fig. 4.16. This would require that very few grains should have the necessary surface energy advantage if, as claimed earlier, growth above the general limit will only occur by the abnormal growth of very few grains into an otherwise unchanged matrix.

If we allow adsorption, or surface oxidation, or the use of substrate (or capping) to reduce the magnitude of the differences in surface energy, then relatively few grains would be able to grow by surface energy effects. A large decrease in surface energy relative to the grain boundary energy, however, would not only reduce the relative surface energy effect but would reduce the thermal groove effect, which would then be at variance with the stagnation of grain growth at $R_0 = 1.5t$.

4.3.4 Secondary Recrystallisation in Foils of Thin Strip

The effects of surface energy described in the previous sections have been discussed in terms of normal grain growth and the stagnation of normal grain growth due to the drag effects associated with thermal grooving. If we consider the situation where all surface energies are equal, ($\Gamma = 1$) and that the surface energy is three times larger than the grain boundary energy, we can see from Fig. 4.16 that in a poly-crystalline array, all grains with diameters less than twice those of its neighbours lie in the stagnant zone and normal grain growth will not occur. In any poly-crystalline array, however, there is a distribution of grain sizes, and it is not uncommon for some grains to have diameters greater than twice the average diameter. Grain size distributions in foils have been assessed (Palmer, Thompson, and Smith, 1987) and an example is shown in Fig. 4.17, for an aluminium foil of thickness t = 30nm. Although this shows a lognormal distribution of diameters, a large majority of grains have diameters in the range 55 – 120nm, (ln(d/t) values from 0.6 – 1.4). Few grains, however, have diameters of 222nm (ln(d/t) values of 2.0) or more. Thus the bulk of the grains will have neighbours whose aggregate grain size will not be greatly different from their own, but there are a few grains with the necessary size advantage (Z = 2 or more) to be in the growth domain, even in the absence of free surface difference effects. These few grains may grow in a static matrix. This phenomenon is known as abnormal grain growth, or secondary re-crystallisation.

The changing nature of the growth process can be illustrated by calculating the borders between stagnation and growth, when the ratio of the matrix grain size to the foil thickness is varied from $R_0/t = 0.5$ up to $R_0/t = 2$. These borders are represented in Fig. 4.18. It should be noted that the model assumes that all structures are entirely composed of through thickness grains. Consider the case, for a given foil thickness, where there is no difference in surface energy due to grain orientation effects, i.e. $\Gamma = 1$. When the diameters of the matrix grains are of the order of the foil thickness ($R_0/t = 0.5$), then all grains with a size advantage above 1.2 (point 'a' in Fig. 4.18) will be able to grow, and as these are likely to be common, 'normal' grain growth will occur and the larger grains will grow, consuming the smaller grains and increasing the average grain size, whilst maintaining a steady state distribution of grain sizes. When the matrix grain diameters reach a size of twice the foil thickness ($R_0/t = 1$), the reduction in driving force is such that a size advantage of Z = 1.5 or more, is required to maintain growth (point 'b'). Thus the proportion of grains that can meet this requirement is reduced but still a significant proportion can grow to change the matrix grain size. This

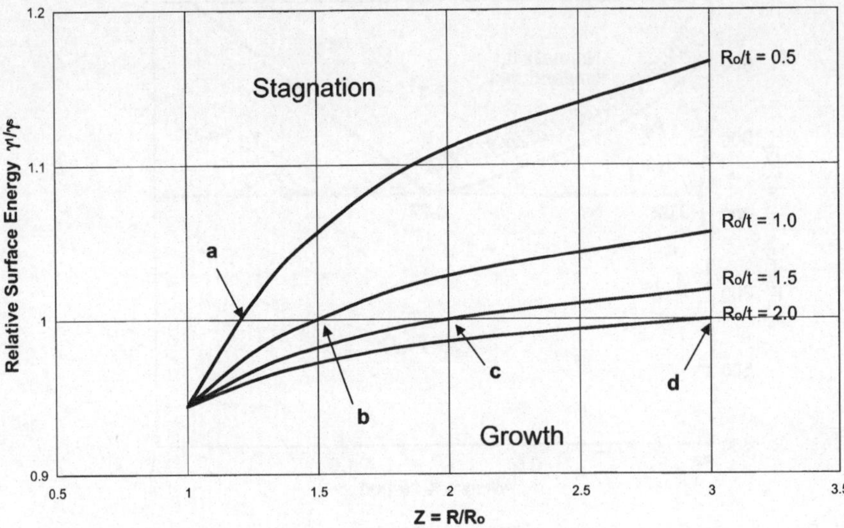

Fig 4.18 Effect of Large Grain Radius on Growth Limit.

process will continue (points 'c' and 'd'), the proportion of grains that will meet the requirement for continued growth will progressively decease until for example at point 'd', the required size advantage for growth is Z = 3. Very few grains may be able to meet this requirement and abnormal growth will occur. If we suppose that only one grain meets this requirement, then only one grain will grow, with the consequence that this grain will grow in an unchanging matrix, i.e. $R_0/t = 2.0$, and the value of Z will be continuously increasing, thus fulfilling the requirements for abnormal grain growth.

If we now consider the case where surface energies do differ, then grains with a reduced surface energy relative to their neighbours will not require such a high size advantage for growth to occur, and therefore a larger proportion of grains should be able to grow. On the other hand there will be instances where grains having a substantial size advantage may be unable to grow because they have an unfavourable surface energy differential. With a bi-variate distribution of size advantage, Z, and surface energy differential, Γ, it would be possible to estimate the proportions of grains that would lie in the growth and stagnation domains of Fig. 4.18 using data such as are given in Figs. 4.15 and 4.17. However, the effects of extrinsic factors on surface energies, and the fact that the surface energy driven growth will result in a significant reduction in surface energy differentials, make such procedures uncertain, but certainly indicate an area for future study.

4.4 THERMAL AND MECHANICAL CYCLES

The maintenance of a fine grain size by inhibiting grain growth pre-supposes that the grain size is fine initially. In promoting a coarse grain size by secondary re-crystallisation it is necessary to include an appropriate thermal treatment where growth is possible. For all fine-grained materials, the thermal and mechanical treatments play an integral part in the development of microstructure. For large cast products, a homogenisation treatment is not

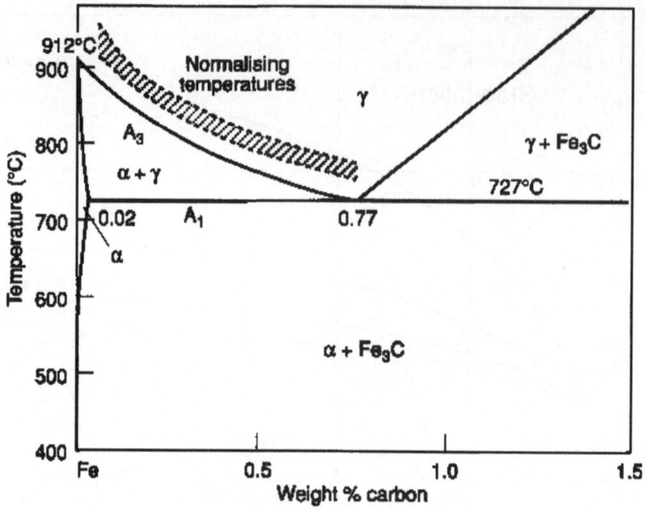

Fig. 4.19 The Iron – Iron Carbide Equilibrium Diagram. The recommended normalising temperatures are shown in relation to the allotropic change points and their dependence on carbon content.

uncommon. For many other products, some form of hot working process is involved, principally to change the shape of the stock material, but increasingly to control the microstructure as well, particularly the grain size. For very thin strip, dimensional precision may demand a cold forming operation, which then may require a subsequent thermal treatment to control the microstructure and restore properties such as ductility and formability. In some cases, the thermal and mechanical treatments are tailored to interact with second phase particles and to control their distribution in the microstructure.

In certain allotropic materials (such as steel for example), a new microstructure is formed each time the stock is heated or cooled through the allotropic change temperatures, and a control of the heating or cooling cycles can be used to develop a fine grained structure.

These topics have provided the subject matter of innumerable books and publications, and only a brief consideration will be given here, mainly in relation to their grain refining effect, and to their interaction with conventional grain refining practices.

4.4.1 Thermal Cycling of Allotropic Materials

As mentioned in Chapter 3, simple thermal cycles through allotropic changes result in one or more complete changes of microstructure, each of which involves a nucleation and growth process. These cycles provide an opportunity to control the microstructures, particularly the grain size. The classic example of using a simple thermal cycle to control the microstructure is the normalising treatment that is applied to steel.

The low carbon end of the Fe – Fe_3C equilibrium diagram is shown in Fig. 4.19. At high temperatures, the austenite (γ) phase is stable, austenite having a f.c.c. structure and showing a high solubility for carbon, as indicated by the line separating the γ and γ + Fe_3C phase fields. At low temperatures, the b.c.c. allotrope of iron, ferrite (α), shows a very limited solubility for carbon. The maximum solubility of carbon in ferrite is shown as 0.02%C at the eutectoid temperature of 727°C. Steels with carbon contents between 0.02% and 0.77% will therefore

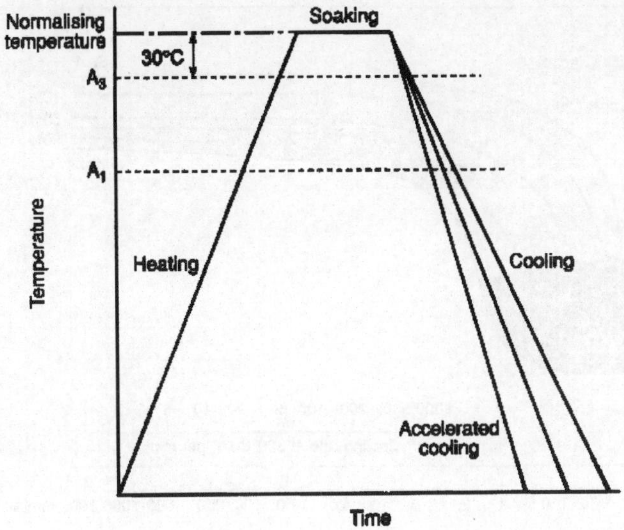

Fig. 4.20 Schematic Normalising Treatment. The heating and cooling rates are influenced by the effective size of the stock, but accelerated cooling rates may be used.

undergo a eutectoid transformation on cooling in which γ with 0.77% transforms to the lamellar eutectoid of cementite (Fe_3C) plates in a ferrite matrix, known as pearlite. The austenite to pearlite transformation will be preceded by the formation of ferrite, the amount of which is controlled by the carbon content of the steel. The amount of ferrite decreases from 100% at 0.02%C to nil at 0.77%C. The microstructures deriving from unspecified hot working processes or from castings can be extremely variable, and in order to provide a more uniform product, a 'normalisation' treatment may be specified. The thermal cycle used in normalising treatments is shown in Fig. 4.20. This involves heating to a normalising temperature, which is 30°C above the upper change point (A_3), soaking for 1 hour per 25mm of equivalent bar diameter, and then air-cooling. The equivalent bar diameter is defined by the effective cooling rate, so that a 25mm equivalent bar diameter relates to a 25mm diameter bar, or a non-circular section that cools at the same rate as the 25mm diameter bar. The equivalent bar diameters can be obtained from heat flow tables, but for plates or strip, the equivalent bar diameter is twice the plate thickness. Although normalising treatments conventionally involve air-cooling of the stock material, accelerated cooling by induced (fan) draft, or by water spray, has been used.

The grain refinement attained in this process is mainly associated with maintaining a fine austenite grain size prior to cooling. This is done by controlling the temperature, as indicated in Fig. 4.19, and in many cases, by adding second phase particles that inhibit grain growth in the austenite temperature range used for normalising. Additional opportunities for grain refinement are associated with the heating and cooling rates during normalising.

A high heating rate gives rise to an increased driving force for the formation of austenite because of the super-heating effect. Under these circumstances the nucleation of austenite grains is increased, and the growth is limited, thus giving a finer austenite grain structure, (Paxton, 1967). There has been some exploitation of rapid re-austenitisation treatments, by electrical resistance heating or by HF induction heating, producing significant austenite

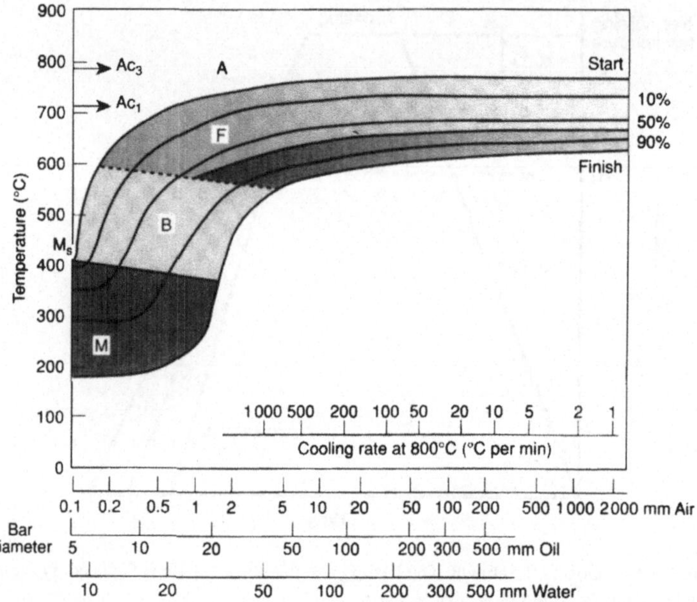

Fig. 4.21 Continuous Cooling Transformation Diagram for a Low Carbon Niobium Steel. The steel was austenitised at 830°C, and the ASTM grain size was ASTM8.

grain refinement, when holding times of only a few seconds are employed. However, such processes are generally restricted to engineering components or bars of limited size.

Increasing the cooling rate from the normalising temperature can affect the austenite decomposition processes, by reducing the temperature of transformation, and refining the transformation products as a result of the increased driving force and the reduced opportunities for ferrite growth during transformation. The continuous cooling transformation diagrams for a variety of steels are well documented, and illustrate the depression of the transformation temperature with increased cooling rate. Conventional continuous cooling diagrams are difficult to use because of the uncertainty of the temperature time path for a given product size. A more convenient form of continuous cooling transformation diagram is shown in Fig. 4.21, (Atkins M 1977) where the transformation temperatures are shown as a function of the equivalent bar diameters. A given point on the abscissa represents the natural cooling of a bar of given diameter. The cooling rate decreases as the stock temperature decreases, so the cooling rate of the stock at 800°C is shown as a scale parallel that of the air-cooled bar diameter. Also the scales for water quenched and oil quenched bars are shown such that their cooling rates correspond to the cooling rates of the equivalent air-cooled bar diameters. For example, a 200mm diameter bar quenched in water will cool at a similar rate to a 170mm diameter bar quenched in oil, which in turn would cool at a similar rate to a 40mm diameter bar cooled in air. The advantage of this form of diagram is that the transformation temperatures are obtained from a vertical intercept of the transformation lines, positioned at the appropriate equivalent bar diameter. The effect of cooling rate on the austenite-ferrite transformation temperature can be illustrated by comparing the 50% transformation temperatures obtained at equivalent air-cooled bar diameters of 500mm and 5mm. The slower cooling of the larger diameter bar gives a

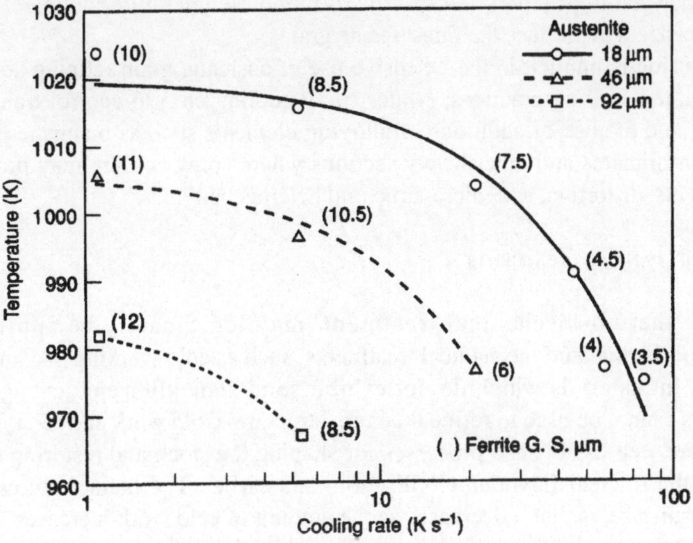

Fig. 4.22 The Effects of Prior Austenite Grain Size on the Ferrite Grain Size in a Low Carbon Aluminium-Treated Steel. (After Militzer et al).

transformation temperature of 700°C, whilst the more rapidly cooled small bar diameter gives a corresponding temperature of 630°C. More rapid cooling than this results in progressive suppression of the polygonal ferrite and the development of increased proportions of bainite and martensite as indicated by the very marked depression of the transformation temperature.

Some indication of the effects of cooling rate on the transformation temperature, and of the effects of both of these factors on the ferrite grain size, is illustrated in Fig. 4.22, (Pandi, Militzer, Hawbolt, and Meadowcroft, 1995), where the effects of cooling rate and the prior austenite grain size on the start of transformation temperature, and on the transformed ferrite grain size, can be seen. The effect of cooling rate on the transformation temperature is readily apparent. It should be noted that these workers used linear cooling from the austenitising temperature and through the transformation, rather than natural cooling, to produce the experimental results. The effect of the prior austenite grain size on the transformation temperature is also quite striking. Increasing the austenite grain size depresses the transformation temperature very markedly, and also reduces the upper limit of the range of cooling rates that can be investigated, due to the suppression of polygonal ferrite and the formation of low temperature transformation products, i.e. bainite and martensite. This is in accord with the well known effect of austenite grain size in increasing the hardenability of steel. There are two effects of the prior austenite grain size on the size of the transformed ferrite grains. Reducing the austenite grain size generally reduces the ferrite grain size at a given cooling rate, although this effect is only modest. The second effect results from the effect of prior austenite grain size on the transformation temperature, and the higher cooling rates that can be used when the austenite grain size is small. Very fine ferrite grains are obtained when high cooling rates are employed in conjunction with a fine austenite grain size (the fine austenite grain size being imperative to the maintenance of a polygonal ferrite micro-structure at very high cooling rates). The major conclusion of this work is that a high degree of under-cooling (depression of transformation temperature) coupled with a fine austenite grain size will produce the finest ferrite grains.

This conclusion is implicit in the extensive use of austenite grain refining additions, the use of accelerated cooling practices, (water spray cooling etc.) to control transformation temperatures, and the use of additions of alloying elements such as manganese to control transformation in plates and other heavy sections where spray cooling may produce large thermal gradients, distortion, and micro-structural heterogeneity.

4.4.2 Thermomechanical treatments

The term 'thermo-mechanical treatment' may legitimately be applied to any combination of thermal and mechanical treatments, such as cold working and annealing, or hot working. In materials which do not exhibit solid-state allotropic changes, simple thermal cycles cannot be used to refine the grain structure. Cold work and re-crystallisation treatments were well-established processes for shaping the stock and restoring the desired properties in the Ancient Egyptian Civilisation, and earlier. The main features of the re-crystallised grain size, is that it decreases as the amount of cold work increases, and also as the grain size prior to deformation is reduced. Increasing the annealing temperature may increase the re-crystallised grain size, but would certainly increase the post re-crystallisation grain growth in the absence of any inhibitors. The cold working and annealing treatments for thin strip, rod, or wire, can be tailored very carefully to produce fine grain sizes, coarse grain sizes, or, when coupled with thermal grooving, atmosphere control, or critical particle dispersions, to produce abnormally huge grains (secondary re-crystallisation) and crystallographic textures.

Such processes are ideal for light sections, where the added dimensional accuracy may provide an additional benefit, but cannot be applied to heavy sections such as thick plate, beams, large diameter bars, etc. With these products, and also as an intermediary process for light products, hot working is used to produce the required shape changes, and, increasingly over the last forty years, to control the microstructure, and hence the mechanical properties. The practice of controlled rolling in ferrous metallurgy expanded due to the improvements of both strength and toughness resulting from refinement of the final ferrite grain size. Indeed, so effective is the control of microstructure and properties obtained by controlled rolling that some product specifications, that had previously required a normalising treatment to give the grain refinement and toughness, were changed, to allow the use of controlled rolled product, thereby eliminating the added cost of a normalising treatment.

The application of controlled processing to an allotropic material such as steel requires that the hot working, carried out in the austenite temperature range, should produce an austenite structure that will, on transformation during cooling and after rolling, produce the desired fine grained ferrite. Most of the controlled rolling is practiced on micro-alloyed steels, where the interplay between the micro-alloy carbides and nitrides and the controlled processing variables can produce interesting effects (Gladman, 1997). Micro-alloying carbides and nitrides may precipitate during hot working and can have two major effects. The most obvious effect is that the presence of the fine particles inhibit grain growth after re-crystallisation, as described earlier in this chapter (Section 4.2.2). The second effect is that re-crystallisation may be severely retarded. This effect is also believed to be associated with strain-induced precipitation. These two effects have led to two distinct controlled rolling processes, i.e. the conventional process, which can produce both re-crystallised and

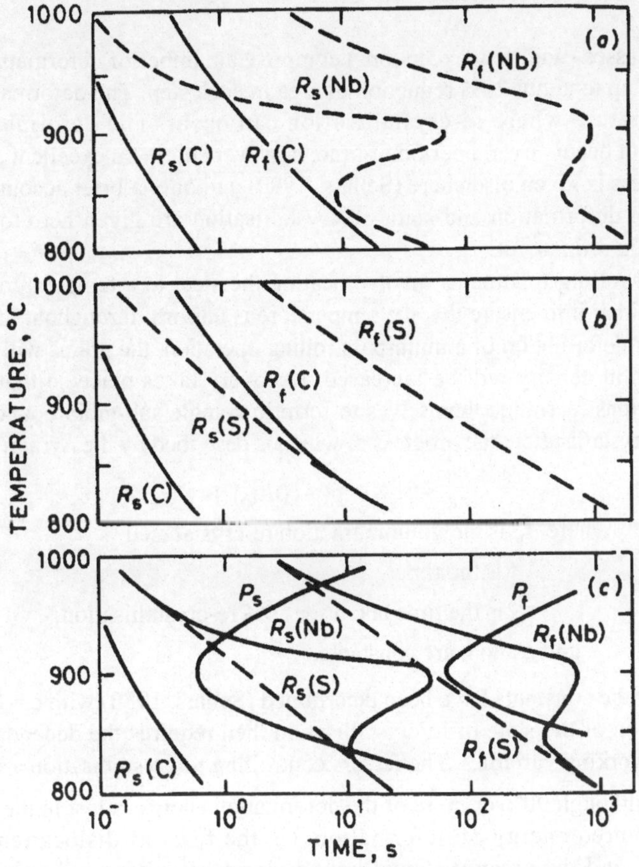

Fig. 4.23 The Effect of Niobium Carbide Precipitation on the Recrystallisation of Austenite. (After Jonas and Weiss).

un-recrystallised austenite grains, and re-crystallisation controlled rolling, which, as its name implies, produces a fully re-crystallised fine austenite grain structure. The effects of micro-alloying on the re-crystallisation of austenite is illustrated in Fig. 4.23(a) (Jonas and Weiss, 1979). When the stock is heated to high temperatures, the micro-alloy carbides or nitrides will dissolve, to a greater or lesser extent (see Chapter 5), but will precipitate as the stock cools during the hot working process. It has been suggested that there may be a solute effect in that dissolved niobium may retard re-crystallisation, Fig. 4.23(b). The C-curve for niobium carbo-nitride precipitation Fig. 4.23(c) causes the added retardation of the deformed austenite. The effects of micro-alloys on the re-crystallisation temperature (Cuddy, 1982) are shown in Fig. 4.24. Niobium has by far the strongest effect in raising the re-crystallisation temperature, whilst vanadium (and aluminium, which is not commonly classed as a micro-alloying element) have the weakest effect. Heavily deformed austenite grains are far more commonly seen in niobium steels because of this effect. Vanadium steels are commonly used in re-crystallisation controlled rolling.

The controlled rolling processes have been described in some detail elsewhere (Gladman, 1997) and only an outline will be given here.

4.4.3 Conventional Controlled Rolling of Steel

Hot rolling processes (with few exceptions) comprise a number of deformation steps, with deformations of up to about 20% reduction of area in each step. The deformation is carried out at a temperature where re-crystallisation can occur. The deformation is never instantaneous, but occurs over a period of time, however short. An excellent account of the hot rolling process is given elsewhere (Sellars, 1980) and only a brief account of the basic principles of hot deformation and static re-crystallisation are given here to illustrate the principles of grain refinement.

Conventional rolling treatments involve heating the steel to a temperature in excess of 1200°C, and soaking it to ensure that the temperature is uniform throughout the stock. If we consider the first deformation of a multi-pass rolling operation, the grains will be elongated, and the dislocation density will be increased. Recovery takes place so that many of the retained dislocations rearrange themselves to form low-angle sub-grain boundaries. This is followed by re-crystallisation, the progress of which is described by the Avrami Equation –

$$f_R = 1 - \exp(-c(t/t_{50})^n) \qquad \text{(equ. 4.50)}$$

where f_R is the volume fraction re-crystallised.

t is the time

t_{50} is the time taken for 50% re-crystallisation,

and c and n are constants.

The values of the constants have been determined (Sellars, 1980) with c = 0.693 and n = 2. A full definition of the state of re-crystallisation then requires the dependance of t_{50} on the various hot working variables. The factors controlling re-crystallisation are –

(a) Strain, ε – although 90% or more of the deformation energy is lost in the form of heat, there is a stored energy of deformation, (in the form of dislocations, sub-grain boundaries, and the extended grain boundaries of the deformed structure), which provides the driving force for re-crystallisation, and which increases with strain.

(b) Temperature, T – re-crystallisation is a thermally activated process, and the rate of re-crystallisation increases with temperature in accord with the Arrhenius relationship and a characteristic activation energy for re-crystallisation, Q_r.

(c) Strain Rate, s – increasing the Zener-Holloman temperature compensated strain rate, S, accelerates re-crystallisation, and is defined by –

$$S = s. \exp Q_d/RT$$

where Q_d is the activation energy for deformation (typically having a value of 213KJ/mole).

R is the gas constant

and T is the absolute temperature, °K.

(d) Prior Grain Size, d_0 – the nucleation of re-crystallised grains occurs mainly, but not exclusively, on prior grain boundaries, so that a refinement of the prior grain size provides more nuclei and therefore accelerates re-crystalisation.

These factors have been incorporated into a general expression describing their effects on t_{50} to give –

$$t_{50} = BS^{m}\varepsilon^{-4} d_0^{2} \exp(Q_r/RT) \qquad \text{(equ. 4.51)}$$

where m is a constant.

Values of m = -0.375 and Q_r = 480KJ/mole have been quoted (Sellars and Whiteman, 1979), although in a later publication Sellars (1980) discounted any effect of temperature compensated strain rate (m = 0) and found $B = 2.5 \times 10^{-19}$s.µm^{-2} and Q = 300 KJ/mole (d_0 being expressed in µm). The value of B appeared to be composition dependent and increased with increasing manganese content in the steel.

Merging Equations 4.50 and 4.51 allows a prediction of the changing state of re-crystallisation with time at any temperature for any given strain and prior grain size. Given that re-crystallisation is allowed to occur between each successive deformation, the change in grain size through the hot working process can be calculated from –

$$d_r = Cd_0^{2/3}\varepsilon^{-1} \qquad \text{(equ. 4.52)}$$

where d_r is the re-crystallised grain diameter
d_0 is the prior grain diameter
ε is the true strain (-ln t/t_0)
and C is a constant (C = 0.5 approximately).

Following re-crystallisation, and in the absence of pinning particles, then grain growth may occur in the time interval between successive deformation according to –

$$d^{n} - d_r^{n} = kt \qquad \text{(equ. 4.53)}$$

where d is the grain diameter after a time t
n is the grain growth exponent
and k is the temperature dependent rate constant.

The rate constant may be expressed as –

$$k = k_0 \exp(-Q_g/RT) \qquad \text{(equ. 4.54)}$$

where k_0 is a constant
and Q_g is the activation energy for grain growth.

Sellars has given values of $k_0 = 4 \times 10^{32}$µm^{10}s^{-1}, and Q_g = 400KJ/mole.

The units of k_0 derive from the experimentally observed values of n = 10 in Equation 4.53. This magnitude of n probably has little significant meaning and merely reflects the empirical nature of the equation when applied to a discontinuous process like grain growth. Such large experimental values of n are not uncommon.

Equations 4.50 to 4.54 are the basic constitutive equations upon which the evolution of microstructure depends during a multi-pass hot rolling process, and are simple to apply when each deformation is followed by complete re-crystallisation, and subsequent grain growth, before the next deformation. If the structure is only partly re-crystallised in the inter-pass interval, then the subsequent deformation will produce another specific deformation in the re-crystallised grains but an accumulated deformation for the unrecrystallised fraction. The accumulated deformation will further accelerate re-

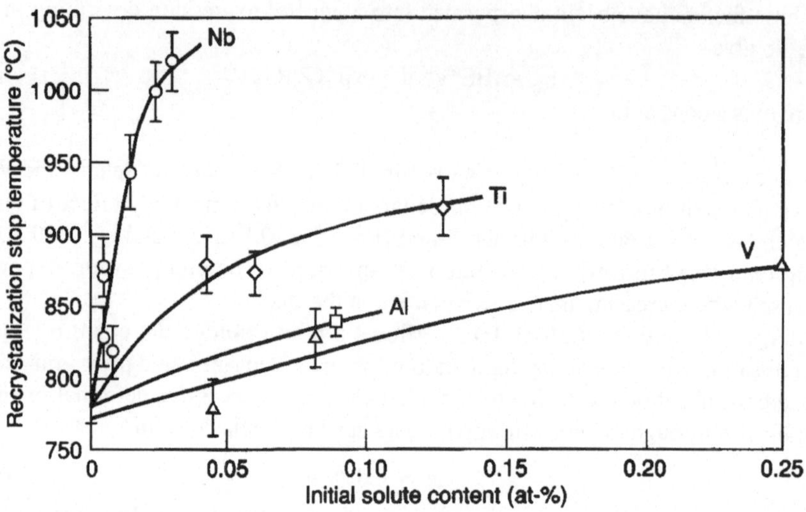

Fig. 4.24 Effect of Micro-alloying Elements on the Recrystallisation Temperature. (After Cuddy).

crystallisation and will provide added grain refinement upon re-crystallisation. In evaluating the grain size in a partially re-crystallised structure, the mechanical reduction of the unrecrystallised grains should be taken into account, particularly if the unre-crystallised grains persist through several deformations (Gladman, 1997, Kozasu, 1992).

Given the basics of the hot working process, the principal methods of controlling the grain size are by reducing the temperature of deformation and by adding second phase particles (grain refining additions), both of which reduce or inhibit grain growth in the interpass and post-rolling periods. Some of the grain refining additives may be so potent that they are able to stabilise the sub-grain boundaries and thereby retard re-crystallisation, particularly when the finish rolling temperatures are low. The latter effect occurs fairly regularly in low temperature controlled rolling, whereas in 'recrystallisation controlled rolling' finishing temperatures are not low and second phase particles are essential to inhibit grain growth (Korchynsky, 1990).

The basic constitutive equations have been used (Sellars, 1980) to predict the evolution of the austenite grain structure during a multipass rolling operation as shown in Fig. 4.25. The rolling schedule involved –

(a) Soaking at 1200°C.

(b) 7 roughing passes of 15% reduction.

(c) Holding to lower the stock temperature.

(d) 8 finishing passes of 15% reduction.

The reduction of grain size from 300μm to 60μm in the carbon-manganese steel shows the recrystallisation and grain growth effects in the roughing stage. The holding period allows the temperature to fall from about 1000°C to 900°C, and the 8 finishing passes are given in the temperature range 900°C to 800°C. In this range, the austenite grain size is reduced from 60μm to 15μm. Adding niobium to steel inhibits grain growth in the

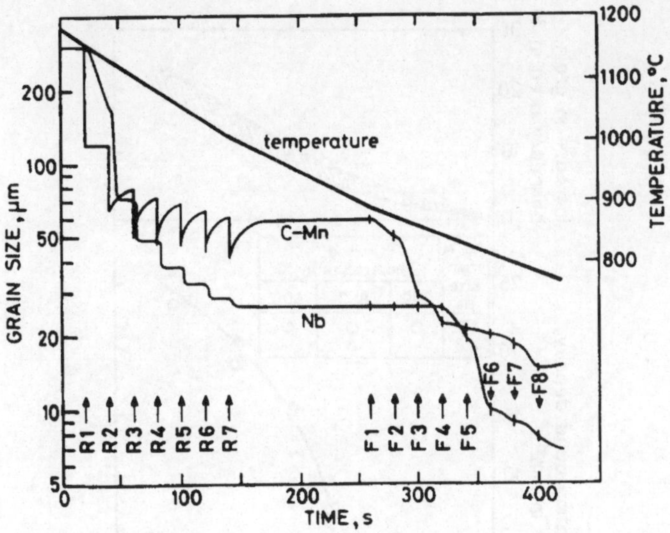

Fig. 4.25 Evolution of Austenite Grain Size During Hot Rolling. (After Sellars).

roughing stage, resulting in a grain size of about 25µm at the holding stage, which is then refined during the finishing passes to about 8µm. Not only will grain growth be inhibited by niobium in the finishing stages, but also the recrystallisation will be severely retarded, as shown in Fig. 4.24, resulting in thin deformed austenite grains, with their attendant internal structures such as sub-grains and shear bands

This presents a fairly simplistic view of the hot rolling process, but has been used successfully to predict the effects of changes in rolling schedules. Far more comprehensive treatments are now available which model the heat flow in the stock material and include the chilling effect of the rolls on the stock surface together with the recovery that occurs in the interpass period, and the heat that is generated by the deformation. As mentioned earlier this can amount to 90% of the energy consumed in rolling. In modern continuous hot bar mills this can result in an increase in temperature from the intermediate stage to the end of rolling. Further, the deformation in the roll pass is far from uniform because of temperature profiles, and the friction effects at the roll/stock interface. Finite element models giving the distribution of strain, strain rates, and temperature are available. The use of these models together with the constitutive equations for microstructural evolution, are currently being developed.

The development of the finite element models to simulate the austenite grain sizes and shapes will need to be extended to cover the ferrite grain size developed from the austenite during cooling. Reference has already been made to work on this subject (Pandi et al, 1995) in connection with normalising treatments. In fact that work was based on as-rolled but recrystallised austenite. Various methods of treating this problem have been tried, (Militzer et al, 1995, Sellars and Beynon, 1985, Hodgson and Gibbs, 1985), but a striking feature of the controlled rolled austenite structures was the effects of heavily elongated austenite grains, and the accompanying internal shear bands, (Kozasu et al, 1977). Fig. 4.26 shows the effects of accumulated deformation below the recrystallisation temperature on the grain boundary area per unit volume, A_G, for a 0.03%Nb steel. Also shown is the effect of the deformation on the deformation band density, A_D. Both of these features act as nuclei for

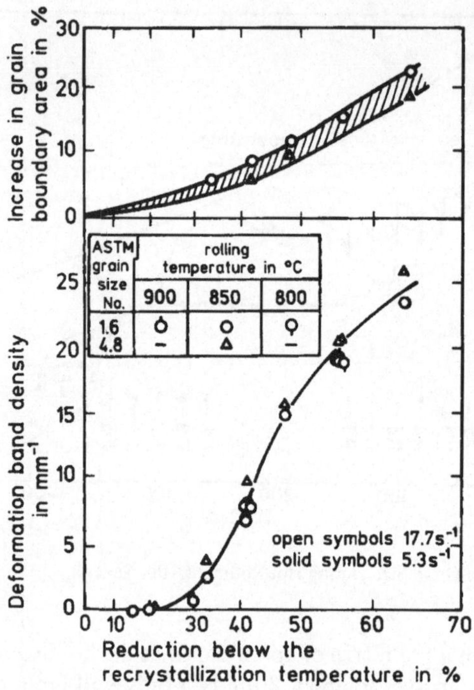

Fig. 4.26 Effect of Low Temperature Deformation on the Grain Boundary Area and Deformation Band Intensity. (After Kozasu, 1992).

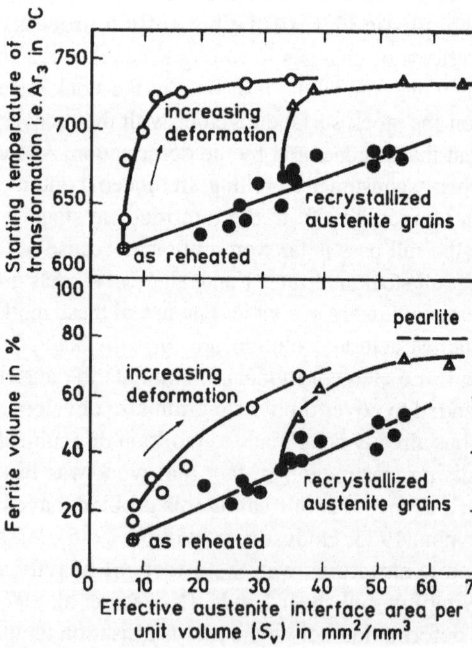

Fig. 4.27 The Effect of Deformation of Austenite on the Allotropic Transformation During Cooling. (After Kozasu et al).

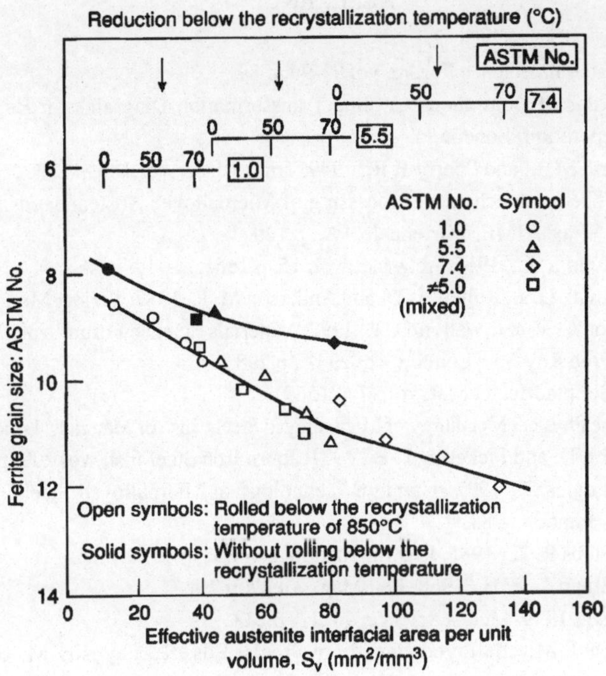

Fig. 4.28 The Effect of Austenite Interfacial Area on the Ferrite Grain Size. (After Kozasu et al).

ferrite during transformation, though not necessarily to the same extent, but the increase in both of these features raises the austenite/ferrite transformation temperature, Fig. 4.27. The overall effect of the increase in nucleation sites is not only to raise the transformation temperature but also to produce a finer ferrite grain size, Fig. 4.28. The grain boundary area per unit volume, A_G, and the area of deformation bands per unit volume were obtained by prescribed stereological methods (Underwood, 1968).

$$A_G = 0.43\, N_p + 1.57 N_n \qquad \text{(equ. 4.55)}$$

where N_p is the number of grain boundaries per unit length measured parallel to the rolling direction,

and N_n is the same measured perpendicular to the rolling direction.

For the deformation bands –

$$A_D = N_L / \sin\theta \qquad \text{(equ. 4.56)}$$

where N_L is the number of deformation band intercepts per unit length measured normal to the rolling plane,

and θ is the angle between the deformation band and the normal to the rolling plane measured in a longitudinal section.

It is clear that a complete finite element model with transformed ferrite grain sizes may yet be some time away, although with a modicum of empiricism, the models are currently possible.

REFERENCES

Aboav D., 1992. Materials Science Forum, vol.94-96, p.89.

Atkins M., 1977. Atlas of Continuous Cooling Transformation Diagrams for Engineering Steels, British Steel Corporation, London.

Beck P. A., Holzworth M. L. and Sperry P. R., 1949. Trans AIME, vol.180, p.163.

Cuddy L. J., 1982. Thermo-mechanical Processing of Microalloyed Austenite, eds. A. J. Deardo, G. A. Ratz and P. J. Wray. AIME, Warrendale, PA., p.129.

Dasarathy C. and Hudd R. C., 1967. Acta Met, vol. 15, p.1665.

Doherty R. D., Srolovitz D. J., Rollett A. D. and Anderson M. P., 1987. Scripta Met, vol.21, p.675.

Frost H. J., Thompson C. V. and Walton D. T., 1992. Materials Science Forum, vol. 94-96, p.543.

Gladman T., 1966. Proc Roy Soc, London, vol.294A, p.298.

Gladman T., 1992. Scripta Met, et Mat, vol.27, p.1569.

Gladman T., 1997. The Physical Metallurgy of Microalloyed Steels, Inst. of Materials, London, (book 615).

Gladman T., McIvor I. D. and Pickering F. B., 1971. Journ Iron Steel Inst., vol.209, p.380.

Gladman T. and Senogles D., 1997. Titanium Technology in Microalloyed Steels, Ed. T. N. Baker, Inst of Materials, London, p.83.

Hodgson P.D. and Gibbs R. K., 1985. ISIJ International, vol.32, p.1329.

Hondros E. D. and Stuart L. E. H., 1968. Phil Mag, vol.17, p.711.

Jonas J. J. and Weiss I., 1979. Metals Science, vol.13, p.238.

Korchynsky M., 1990. Microalloyed Vanadium Steels, Eds. Korchynshy M. Gorczyca S. and Blicharski M., Assn Polish Met Engrs, Cracow.

Kozasu I., 1992. Materials Science and Technology, vol.7, Constitution and Properties of Steels, Ed. F. B. Pickering, V.C.H., Weinheim, Germany, p.183.

Kozasu I., Ouchi C., Sampei I. And Okita T., 1977. Microalloying 75, Union Carbide, New York, p.120.

Martin J. W. and Doherty R. D., 1976. Stability of Microstructure in Metallic Systems, Cambridge University Press, Cambridge, U.K.

Millitzer M., Hawbolt E. B. and Meadowcroft T. R., 1995. HSLA Steels 1995, Eds. G. Liu, H. Stuart, H. Zhang and C. Li, Chinese Society for Metals, Beijing, p.271.

Mullins W. W., 1958. Acta Met, vol.6, p.414.

Mullins W. W., 1963. Metal Surfaces, ASM, Metals Park, Ohio, p.17.

Palmer J. E., Thompson. C. V. and Smith H. I., 1987. Journ Applied Phys, vol.62, p.2492.

Pandi R., Millitzer. M., Hawbolt E. B. and Meadowcroft T. R., 1995. HSLA Steels 1995, Eds. G. Liu, H. Stuart, H. Zhang and C. Li, Chinese Society for Metals, Beijing, p.265.

Paxton H., 1967. Transformation and Hardenability in Steel, Climax Molibdenum Co., Ann Arbor, Michigan, p.3.

Sellars C. M., 1980. Hot Working and Forming Processes, Eds. C. M. Sellars and G. J. Davies, book No. 264, The Metals Society, London, p.3.

Sellars C. M. and Beynon J., 1985. High Strength, Low Alloy Steels, 85, Eds. D. Dunne and T. Chandra, South Coast Printers, p.142.

Senogles D., 1994. PhD. Thesis, School of Materials, University of Leeds.

Srolovitz D. J., Anderson M. P., Grest G. S. and Sanhi P. S., 1984. Acta Met, vol.32, p.1429.

Sundquist E., 1964. Acta Met, vol.12, p.67.

Thompson C.V., 1992. Materials Science Forum, vol.94-96, p.245.

Underwood E. E., 1968. Quantitative Microscopy, Eds. R. T. DeHoff and F. N. Rhines, McGraw-Hill, New York, p.77.

Zener C., 1948. See discussion in C. S. Smith, Trans AIMME, vol.175, p.47.

Chapter 4 Exercises

1. Consider an array of grains where the matrix (or average) grain radius is 10μm. What volume fraction of randomly distributed particles, with a uniform radius of r = 10nm would be required to inhibit the growth of a grain with a size advantage (Z) of 5/3?

2. Given the same parameters as in Exercise 1, what volume fraction would be required if all of the particles were on the grain boundaries?

3. Given the same parameters as in Exercise 1, with randomly distributed particles, but with a larger grain of Z = 2 in the array, what volume fraction of particles would inhibit the growth of this grain?

4. If a volume fraction of 2×10^{-4} of randomly distributed particles of radius 10nm, were added to a material already containing a volume fraction of 2 x 10 ^{4} of randomly distributed particles of 20nm radius, what matrix grain size could be stabilised when large grains have a size advantage of Z = 5/3?

CHAPTER 5

Second Phase Particle Contents and Particle Coarsening

5.1 INTRODUCTION

Second phase particles occur as a result of their thermodynamic stability, and their distribution and morphologies may also be affected by the thermal and thermomechanical treatments given, and by the alloy design. As described in the previous chapter, the volume fraction of particles, their location, and the size of the particles, are important with respect to the control of grain size. In this chapter, therefore, the thermodynamic and alloy design aspects will be considered, and a brief description of the formation of the second phase particles will be given. Because of the importance of particle size, the particle growth process, known generally as Ostwald ripening, will also be considered. The type of particle, and the alloy chemistry, can have important effects on the stability of the particles at elevated temperatures, and on the particle coarsening rates, as will be shown.

In a bulk engineering product such as steel, the cost of the grain refining additions can also be very important, and this does not merely relate to the cost of the alloying elements themselves. For example, prior to 1950, aluminium was the main grain refining additive. Now aluminium itself is not particularly expensive, especially when added in amounts of about 0.03wt.%. However aluminium has a high affinity for oxygen, so that aluminium additions could only be made to a well deoxidised steel. Well deoxidised steels suffer extensive solidification shrinkage and need hot topped ingots to feed liquid metal into the shrinkage cavity. This results in a product yield of only 75% of the liquid mass. Balanced steels on the other hand, have yields of 95%, or thereabouts, because the shrinkage is balanced by carbon dioxide gas bubbles formed in mildly deoxidised steel during solidification. The cost of grain refinement by aluminium treatment was therefore very high. In the late 1950's attention turned to the use of niobium and vanadium as grain refiners. Both of these elements have a low affinity for oxygen and could therefore be used in balanced steels. Ironically, the widespread adoption of the continuous casting process for bulk steel manufacture, from the 1980's onwards, required a well deoxidised steel, but by virtue of the process change, could deliver a 98% yield of billet, bar, or slab. Elements with a high affinity for oxygen, such as aluminium and titanium, could therefore be used, without the punitive costs associated with high yield losses.

The effects of these grain-refining additions depend upon the formation of nitrides, carbides, or carbo-nitrides. Aluminium in particular exerts its grain refining influence by the formation of aluminium nitride. In the middle of the twentieth century, steel making was carried out by the open hearth, basic electric, and Bessemer steel making processes, the characteristic nitrogen contents being 0.008wt.%, 0.011wt.%, and 0.015wt.%

respectively. Despite the variation, there was sufficient nitrogen for the usual aluminium additions to produce enough aluminium nitride to grain refine the steel. The development of the Basic Oxygen Steel making (BOS) process in the latter part of the twentieth century, however, resulted in nitrogen contents below 0.005wt.%, so that steels with aluminium or vanadium now often require additional nitrogen.

It can thus be appreciated that the practice of grain refinement may be influenced by external factors such as process changes, alloy pricing policies, etc., but it is hoped that the basic principles of grain refining additives outlined in this chapter will provide the flexibility to meet such changes.

5.2 PARTICLE SOLUBILITY

Alloy systems that will give a particulate distribution of the second constituent or phase, often show a terminal solid solution having a limited solubility of the second constituent, where the solubility decreases as the temperature is lowered. A classic example is the binary Al-Si system, which is a simple eutectic system. The aluminium rich terminal solid solution can contain up to about 1.2wt.% silicon at the eutectic temperature of 577°C, but this value decreases rapidly with decreasing temperature. The second constituent phase in the aluminium rich alloys is almost pure silicon. Alloys with 1.2wt.% Si or less will solidify to give a single f.c.c. phase (α), i.e. a solid solution of silicon in aluminium. On cooling below the eutectic temperature, silicon (diamond cubic structure) is precipitated, because the solubility of silicon in the α-phase decreases. The silicon particles are generally spheroidal.

There are many instances where the second phase precipitated from solution is not elemental, but is an intermetallic or other compound. Aluminium rich Al-Cu alloys will precipitate the θ phase, $CuAl_3$, the mechanism of which provides the basis for precipitation hardening alloys such as duralumin. In steels with less than 0.02wt.%C, particles of cementite (Fe_3C) are precipitated from the b.c.c. ferrite solid solution at temperatures below the eutectoid temperature (727°C). However, these precipitation systems have not been developed for their grain refining propensities. The grain refining particles in steel, for example, have much higher thermodynamic stabilities and can only be dissolved at relatively high temperatures. Their resistance to coarsening is also much greater. These constituents are often alloy carbides and nitrides, and in order to understand the factors that control the particle characteristics, we must examine their solubility and the effects of alloy chemistry in these ternary (or higher) systems.

5.2.1 The Solubility Product

The components used to create grain refining particles may be present either in solution or combined to form the constituent particle. There may be one or more constituents (types of particle) some of which may have a common component. For example, aluminium and nitrogen (the components) in steel may react to form aluminium nitride (the constituent phase). If titanium is then added, a second constituent phase titanium nitride may be formed, thus providing two constituents (aluminium nitride and titanium nitride) with a common component (nitrogen). The addition of a component, e.g. carbon, may cause the titanium nitride to change its composition to become a carbo-nitride, i.e. there is an added component, but there is no extra constituent, merely a chemical change to an existing constituent. These changes in alloy composition have significant effects on the matrix composition, and on the stability and composition of the second phase particles, but in order to understand them, it is first necessary to understand the solubility product of a simple carbide or nitride.

Consider a reaction between two solutes, M and X, in a steel, to form a compound MX.

$$[M] + [X] \rightarrow (MX) \qquad \text{(equ. 5.1)}$$

where [M] is the component, M, dissolved in the steel
[X] is the component, X, dissolved in the steel
and (MX) signifies the formation of a separate constituent (phase).

The equilibrium constant K, is traditionally defined as –

$$K = a_{(MX)}/a_{[M]}.a_{[X]} \qquad \text{(equ. 5.2)}$$

where $a_{(MX)}$ is the activity of the constituent MX
$a_{[M]}$ is the activity of the solute M
and $a_{[X]}$ is the activity of the solute X.

The solubility product, k_s, is defined as the reciprocal of the equilibrium constant, K, so that –

$$k_s = a_{[M]}.a_{[X]}/a_{(MX)} \qquad \text{(equ. 5.3)}$$

The activities of the solute in Equation 5.3 can be expressed in terms of the elemental concentrations and activity coefficients, γ, [M] and [X],

$$a_{[M]} = [M].\gamma_{[M]}$$
$$a_{[X]} = [X]\ \gamma_{[X]}$$

to give –

$$k_s = [M].\gamma_{[M]}\ .\ [X].\gamma_{[X]}/a_{(MX)} \qquad \text{(equ. 5.4)}$$

In a very empirical manner, and with all the aplomb of a late 20th Century practicing metallurgist, it is assumed that the activity coefficients $\gamma_{[M]}$ and $\gamma_{[X]}$ are equal to unity, and that the activity of the pure product MX is of course unity, so that Equation 5.4 may be written as –

$$k_s = [M]\ .\ [X] \qquad \text{(equ. 5.5)}$$

Strictly speaking, [M] and [X] should be the molar fractions of solute, but, particularly at the low concentrations used in grain refined steels (typically less than 0.1wt.%) the use of elemental concentrations expressed in terms of percentage by weight merely alters the value of k_s. This is of little significance provided that it is understood that k_s has been defined in this way.

For compounds with different atomic ratios, for example M_4X_3, the reaction is –

$$4[M] + 3[X] \rightarrow (M_4X_3) \qquad \text{(equ. 5.6)}$$

The solubility may be defined in different ways, depending upon whether it relates to 1 mole of M_4X_3, 1 mole of M, or 1 mole of X. For Equation 5.6, using 1 mole of M_4X_3 as the basis –

$$k_s = [M]^4\ [X]^3 \qquad \text{(equ. 5.7)}$$

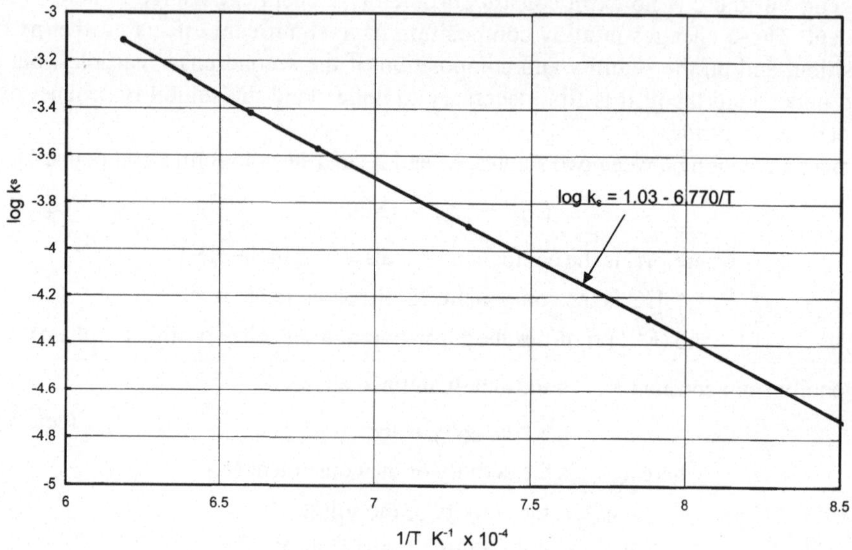

Fig 5.1 Solubility Product for Aluminium Nitride.

If however Equation 5.6 is rewritten as –

$$4[M]/3 + [X] \rightarrow (M_4X_3)/3 \qquad \text{(equ. 5.8)}$$

then the solubility product (based on 1 mole of X), becomes –

$$k_s = [M]^{4/3}\,[X] \qquad \text{(equ. 5.9)}$$

The solubility products expressed in Equations 5.8 and 5.9 are clearly different, the former being the cube of the other. Either of these may be used, again with the proviso that the definition is specified.

The temperature dependence of the solubility product, k_s, can be expressed by the Arrhenius relationship –

$$\ln k_s = (-\,Q/RT) + c \qquad \text{(equ. 5.10)}$$

where Q is the heat of dissolution

R is the gas constant

T is the absolute temperature (°K)

and c is a constant.

Historically, Equation 5.10 has been transformed to give $\log_{10} k_s$, i.e.

$$\log_{10} k_s = (-\,Q/2.303RT) + A \qquad \text{(equ. 5.11)}$$

where $A = c/2.303$

The results are often expressed as –

$$\log k_s = A - B/T \qquad \text{(equ. 5.12)}$$

where $\log k_s$ is the logarithm to the base 10

k_s is defined with [M] and [X] in weight percentages
T is the thermodynamic temperature (°K)
and B = Q/2.303R

This definition of the solubility product and its temperature dependence is used throughout this book, unless otherwise specified.

5.2.2 Using Simple Solubility Products

Given the constants A and B for a given compound, MX, the solubility product for MX in the matrix of a ternary Y-M-X alloy can be evaluated at any specified temperature. If we take the Fe-Al-N system as an example, aluminium in solution in the γ-iron matrix can react with nitrogen in solution in the matrix to form particles of aluminium nitride.

$$[Al] + [N] \rightarrow (Al\,N) \qquad \text{(equ. 5.13)}$$

The solubility product for aluminium nitride in austenite (the f.c.c. form of iron) is given (Irvine, Pickering and Gladman, 1967) as –

$$\log k_s = -6{,}770/T + 1.03 \qquad \text{(equ. 5.14)}$$

Fig. 5.1 shows this relationship, where log k_s is plotted as a function of reciprocal temperature (1/T). Over the range of temperatures 900 to 1400°C (1173 to 1673°K), appropriate to the austenite temperature range of low carbon structural steels, the value of k_s varies from 1.8×10^{-5} to 9.6×10^{-4}. Values of k_s at various temperatures in this range are shown in Table 5.1. If we select a temperature of, for example, 1000°C, the solubility product is given as 5.15×10^{-5}. From the definition of the solubility product, we can write –

$$[Al]\,[N] = 5.15 \times 10^{-5} \qquad \text{(equ. 5.15)}$$

This means that under equilibrium conditions, the product of the dissolved aluminium and dissolved nitrogen contents in the austenite solid solution will be equal to the solubility product. Re-arranging Equation 5.15, we have –

$$[Al] = 5.15 \times 10^{-5} / [N] \qquad \text{(equ. 5.16)}$$

Table 5.1 Effect of Temperature on the Solubility Product of Aluminium Nitride.

Temperature °C	Temperature °K	log k_s	k_s
900	1173	-4.74	1.82×10^{-5}
950	1223	-4.51	3.12×10^{-5}
1000	1273	-4.29	5.15×10^{-5}
1050	1323	-4.09	8.18×10^{-5}
1100	1373	-3.90	12.60×10^{-5}
1150	1423	-3.73	18.62×10^{-5}
1200	1473	-3.57	27.20×10^{-5}
1250	1523	-3.42	38.40×10^{-5}
1300	1573	-3.27	53.20×10^{-5}
1350	1623	-3.14	72.20×10^{-5}

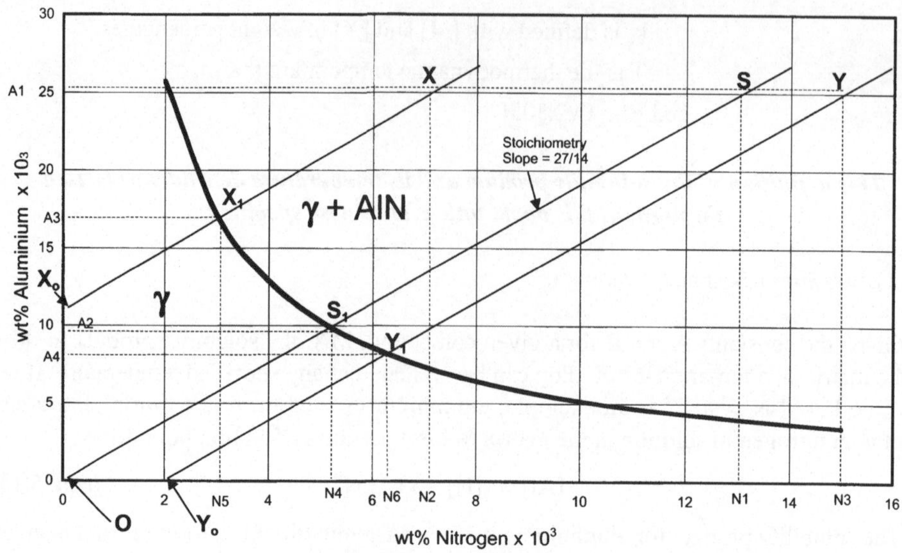

Fig 5.2 Solubility of Aluminium Nitride at 1000°C.

This describes the relationship between the maximum dissolved aluminium content, at the selected temperature of 1000°C, that can exist under equilibrium conditions for a given dissolved nitrogen content. If the dissolved nitrogen content is 0.001wt.%, then the maximum dissolved aluminium content would be 5.15 x 10^{-2} or 0.0515wt.%. If the aluminium content is 0.01wt.%, then the maximum dissolved aluminium content would be 5.15 x 10^{-3} or 0.00515wt.%. If the aluminium content of the alloy exceeds the maximum dissolved value, (this varying with dissolved nitrogen content), then precipitation of aluminium nitride would occur in order to lower both the dissolved aluminium and nitrogen contents, until they complied with the solubility product. Under these circumstances the solid solution would co-exist with the second phase, in this case - aluminium nitride. If the aluminium content of the alloy was less than that permitted by Equation 5.16, then both solutes would co-exist in solid solution, and aluminium nitride would not be precipitated. Thus, Equation 5.16 constitutes a boundary in the Fe-Al-N ternary system, at 1000°C, between the single-phase (austenite) domain, and the two-phase (austenite plus aluminium nitride) domain. This phase boundary is depicted in Fig. 5.2, where the single phase (γ) and two-phase (γ + AlN) domains are clearly shown.

If we consider the reaction between aluminium and nitrogen to form aluminium nitride, as shown in Equation 5.13, then one aluminium atom of relative atomic mass $A_{Al} = 27$, will react with one nitrogen atom of relative atomic mass $A_N = 14$, to give one mole of AlN with a relative molecular mass of 41, i.e. –

$$27\text{gm. of Al} + 14\text{gm. of N} \rightarrow 41\text{gm. of AlN},$$

i.e. the ratio of the mass of aluminium to the mass of nitrogen reacting to form aluminium nitride is constant at 27:14. This is known as the stoichiometric ratio.

If we define a stoichiometric alloy as one whose weight percentages of aluminium and nitrogen are in stoichiometric ratio, then the composition of these alloys will lie on the line

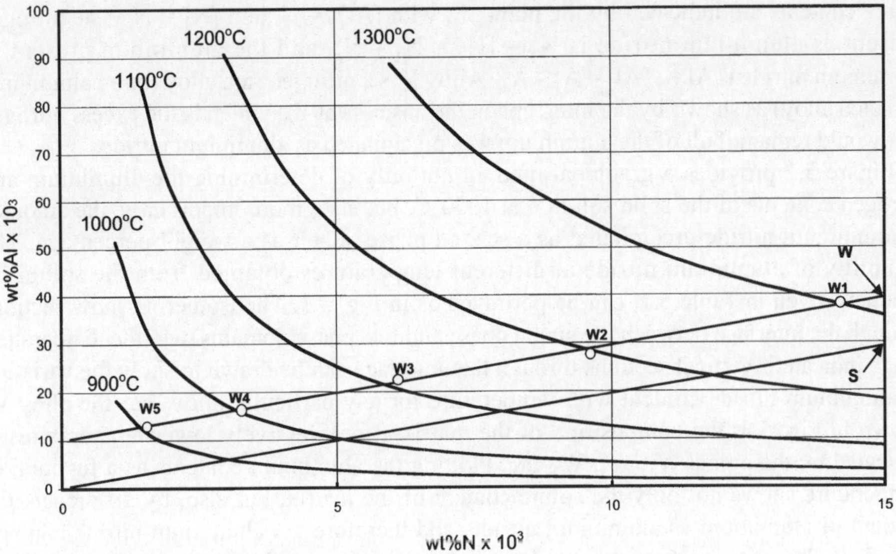

Fig 5.3 Solubility Isotherms for Aluminium Nitride.

OS shown in Fig. 5.2. They can occur in either the single phase or two phase domains. In the two phase state (e.g. an alloy of composition S), because the two elements always react in stiochiometric amounts, the solid solution which is left after precipitation will also contain stoichiometric amounts of aluminium and nitrogen, as depicted by the point S_1 in Fig. 5.2. In stoichiometric alloys having compositions lying along the line S-S_1, therefore, the precipitation of aluminium nitride (at 1000°C) will result in an equilibrium solid solution of composition S_1. In an alloy of composition S, having N_1wt.% nitrogen and A_1wt.% aluminium, equilibration at 1000°C will produce aluminium nitride, and leave a solid solution of composition S_1, having [N] = N_4wt.% nitrogen and [Al] = A_2wt.% aluminium. In the alloy S, the total nitrogen content, N_T is shown as N_1, and the total aluminium content, Al_T, is shown as A_1. The amount of nitrogen present as aluminium nitride, therefore, is (N_T – [N] wt.% or from Fig. 5.2, N_1- N_4. The amount of aluminium present as aluminium nitride is (Al_T – [Al])wt.%, or from Fig. 5.2, ($A_1 - A_2$)wt,%. Thus the amount of aluminium nitride present is ($A_1 - A_2 + N_1 - N_4$)wt.%.

It will be clear, however, from Fig. 5.2 that most alloys will not have stoichiometric compositions. Consider for example the two alloy compositions denoted by points X and Y. Point X shows an alloy which has a deficiency of nitrogen (or a surplus of aluminium) and if all of the nitrogen could be precipitated as aluminium nitride the amount of aluminium left in solution is shown by the point X_0. By definition, the amount of aluminium required to react with nitrogen to form aluminium nitride is always 27:14, so that progressive precipitation of aluminium nitride causes the composition of the solid solution to move along the line X-X_0 parallel to the stoichiometric ratio shown by the line SO. Equilibrium at 1000°C would give a solid solution of composition X_1 with [Al] = A_3 and [N] = N_5. This composition clearly does not have stoichiometric aluminium and nitrogen contents. The amount of aluminium nitride is obtained from the aluminium and nitrogen present as aluminium nitride, ($A_1 - A_3$) and ($N_2 - N_5$)wt.% respectively. Similarly, for the alloy Y, the solute contents are indicated by the point Y_1, with [Al] = A_4, and [N] = N_6. The nitrogen present as aluminium nitride is N_T – [N] = $N_3 - N_6$, and the aluminium present as aluminium nitride is Al_T – [Al] = $A_1 - A_4$. Alloy Y is a nitrogen rich alloy (or an aluminium depleted alloy) as shown by the intercept on the abscissa at Y_0, which is the excess nitrogen that would remain if all of the aluminium was precipitated as aluminium nitride.

Figure 5.2 provides a graphical method not only of determining the aluminium and nitrogen contents of the solid solution at 1000°C, but also, more importantly, the amounts of aluminium nitride precipitated as a second phase, albeit as a weight percentage. The solubility of aluminium nitride at different temperatures obtained from the solubility products given in Table 5.1, can be portrayed as in Fig. 5.3. The isotherms show sections through the interface between the single phase and two phase domains over the temperature range indicated. Vertical sections through this interface can be drawn to show the variation in aluminium nitride content with temperature for any particular alloy. For the alloy W, shown in Fig. 5.3, the composition of the matrix at progressively lower temperatures is indicated by the points W_1, W_2, W_3, etc. Plotting the aluminium contents as a function of temperature shows not only the solute content of the matrix, but also, by difference, the amount of aluminium as aluminium nitride, and therefore the aluminium nitride content, Fig. 5.4. The extent of the phase domains will of course be dependent on both the aluminium and nitrogen contents of the alloy, and in particular, the departure from stoichiometry, as will be appreciated from Fig. 5.2.

5.2.3 Calculating the Particle Content for Simple Compounds

The graphical method of estimating the amount of second phase given in the previous section is clearly based on very simple concepts capable of being treated to give a mathematical expression as follows.

The total nitrogen content, N_T, is equal to the sum of the dissolved nitrogen content [N], and that which is present as the nitride N_{AlN}. Similarly, the total aluminium content, Al_T, is equal to the sum of the dissolved aluminium content [Al] and that present as the nitride Al_{AlN}.

$$N_T = [N] + N_{AlN} \qquad \text{(equ. 5.17)}$$

$$Al_T = [Al] + Al_{AlN} \qquad \text{(equ. 5.18)}$$

$$\text{Now, } [N][Al] = k_s \qquad \text{(equ. 5.19)}$$

where k_s is the solubility product for the appropriate temperature. Transposing from Equation 5.17 and 5.18 gives –

$$(N_T - N_{AlN})(Al_T - Al_{AlN}) = k_s \qquad \text{(equ. 5.20)}$$

From the stoichiometric ratio, we know that –

$$N_{AlN} = 14(Al_{AlN})/27 \qquad \text{(equ. 5.21)}$$

Substituting this for N_{AlN} in Equation 5.20 gives –

$$(N_T - 14Al_{AlN}/27)(Al_T - Al_{AlN}) = k_s \qquad \text{(equ. 5.22)}$$

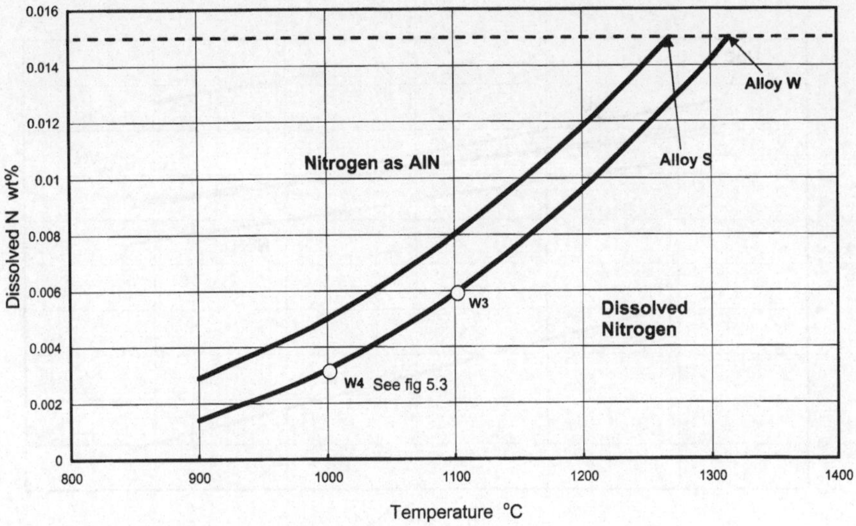

Fig 5.4 Effect of Temperature on Dissolved Nitrogen.

Expansion gives a quadratic equation of the form –

$$a(Al_{AlN})^2 + b(Al_{AlN}) + c = 0 \qquad \text{(equ. 5.23)}$$

where $a = 14/27$

$b = -14Al_T/27 - N_T$

$c = N_TAl_T - k_s$

The solution to the quadratic equation is given by –

$$Al_{AlN} = (-b \pm \sqrt{(b^2 - 4ac)})/2a \qquad \text{(equ. 5.24)}$$

so that the aluminium present as aluminium nitride can be calculated by substituting the appropriate values for a, b, and c. In other words, given the total aluminium and nitrogen contents of a steel and a knowledge of the solubility product, the extent of aluminium present as aluminium nitride can be calculated. From stoichiometrics, (Equation 5.21), the nitrogen as aluminium nitride (N_{AlN}), can be calculated, and the dissolved nitrogen [N] and aluminium [Al] can be obtained from Equations 5.17 and 5.18 respectively.

It should be noted that this assumes that the aluminium and nitrogen is present either in solution or as aluminium nitride. If, for example, there was a high oxygen content in the steel, some of the aluminium would be present as very stable alumina (Al_2O_3), and the aluminium present in the oxide would need to be subtracted from Al_T in Equation 5.18 before following the calculation procedure.

5.2.4 Solubility Data for Other Grain Refining Additives in Steel

The late 1950's saw the introduction of a new class of steels known as micro-alloyed steels. The reasons for their introduction have been described elsewhere (Gladman, 1997), but an important feature of their success was their fine grain size, created either during processing stages, such as

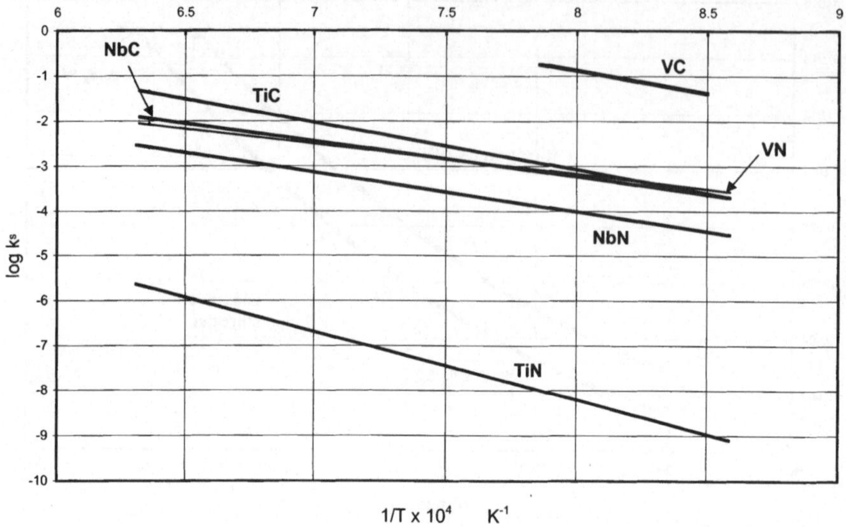

Fig 5.5 Carbide and Nitride Solubilities.

hot rolling, or in subsequent thermal treatments. The particular elements involved include vanadium, niobium, and titanium, all of which form carbides and nitrides in steel. A complication which arises when these alloys are used, is that their carbide and nitrides all have the same cubic crystal structure and show very similar lattice parameters, so that these compounds show extensive mutual solid solubility, i.e. they can form complex carbo-nitrides. This will be discussed in the next section, but first we will consider the individual carbide and nitride solubilities.

The data for the solubility product, expressed in logarithmic form as in Equation 5.12, are shown in Table 5.2. The table shows only one relationship for each carbide or nitride, and only the solubility in the austenite phase is given. This material is condensed from a much larger body of work, (see Gladman, 1997). The data can be used to establish Arrhenius type graphs, showing the logarithm of the solubility product as a function of reciprocal temperature, similar to that for aluminium nitride shown in Figure 5.1. Fig. 5.5 summarises the solubilities of the

Table 5.2 Solubility Product Data for Micro-alloy Carbides and Nitrides in Austenite.

Carbides	**A**	**B**	**Reference**
Vanadium Carbide	6.72	9,500	Narita, 1975
Niobium Carbide	3.42	7,900	Narita, 1975
Titanium Carbide	5.33	10,475	Narita, 1975
Nitrides			
Vanadium Nitride	3.02	7,840	Roberts et al, 1985
Niobium Nitride	2.80	8,500	Narita, 1975
Titanium Nitride	3.82	15,020	Narita, 1975
Aluminium Nitride	1.03	6,770	Leslie et al, 1954

Note: $\log k_s = A - B/T$ (°K)

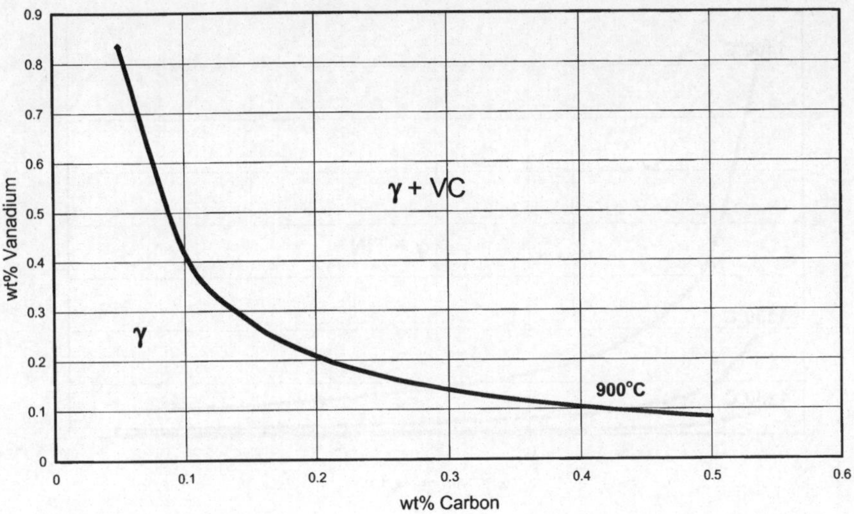

Fig 5.6 Solubility of VC in Austenite at 900°C.

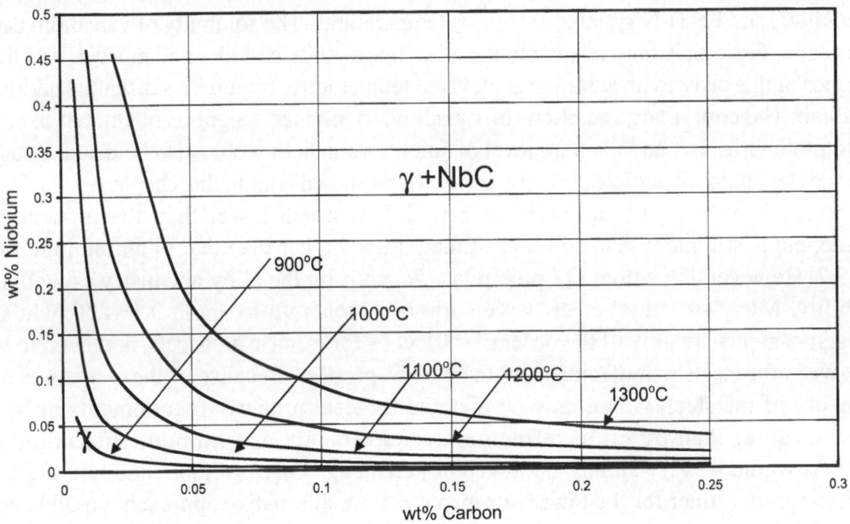

Fig 5.7 Solubility of Niobium Carbide in Austenite.

carbides and nitrides in austenite. For each of the micro-alloying elements, the nitride is less soluble than the carbide. Several of the carbides and nitrides show similar solubilities in the austenite, but with two notable exceptions. Vanadium carbide is far more soluble in austenite, and titanium nitride far less soluble, than other carbides and nitrides. The solubilities change by several orders of magnitude, thus providing a wide range of particle stabilities, each of which may be applied to different problems, as will be shown in Chapter 6.

Some idea of the significance of the differences between these solubility products can be

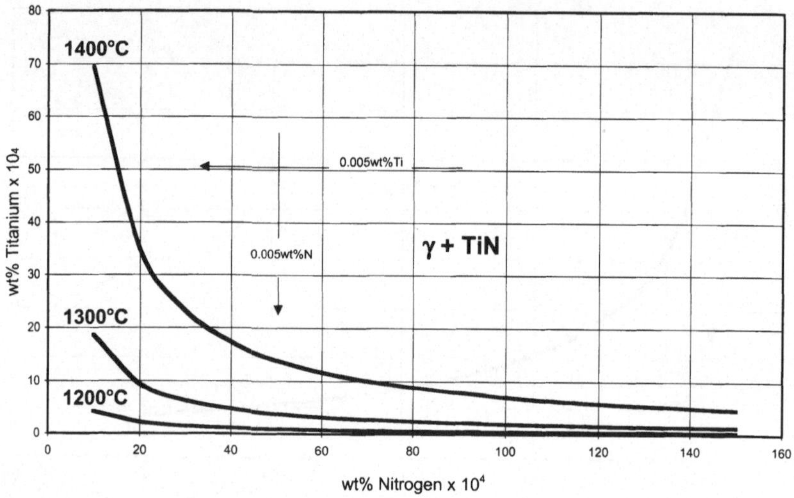

Fig 5.8 Solubility of TiN in Austenite.

obtained from Fig. 5.6, 5.7, and 5.8, which show the austenite phase boundaries for the Fe-V-C, Fe-Nb-C, and Fe-Ti-N systems at various temperatures. The solubility of vanadium carbide in austenite, Fig. 5.6, is high relative to the other micro-alloy carbides and nitrides, and the co-existence of this phase with austenite at elevated temperatures requires a substantial addition of vanadium. The cost of large additions of vanadium to produce a significant amount of carbide would prohibit its use, and the high level of soluble vanadium would also be disadvantageous from the standpoint of particle coarsening, as will be shown later in this chapter.

The solubility of niobium carbide, Fig. 5.7, is much lower than that of vanadium carbide, but is still more than an order of magnitude higher than that of aluminium nitride, Fig. 5.7. However, the amount of precipitate depends on the alloy chemistry as well as the solubility. Many structural steels have carbon contents up to about 0.1wt.%, whilst the nitrogen contents are only of the order of 0.01wt.% for electric arc steels, and even lower at 0.004wt.% for steels manufactured by the B.O.S. process. Because of these features of the chemistry of the steel, which may be dictated by steel property requirements or by cost aspects, rather than by grain refinement requirements, the niobium and aluminium additions would be very similar on a weight percentage basis. In fact, niobium might be a preferred grain refiner for the lower nitrogen steel. An alternative approach would be to use a nitrogen enhancing ferro-alloy so that the potency of aluminium nitride would be restored. The cost of the additives and their effects on processing response and ancillary properties would probably be decisive, but these factors illustrate the flexibility that is offered by a detailed knowledge of the grain refining principles and the solubility data.

Titanium nitride, Fig. 5.8, shows a far lower solubility than any of the other micro-alloy carbides and nitrides, and also lower than that of aluminium nitride. In fact, the high stability of titanium nitride is such that the bulk of the nitride will be precipitated either in the liquid state or during solidification, thus producing very large cubic particles, some several microns in edge length in bulk materials. These are not very useful for grain refinement as indicated by the direct relationship between particle size and stabilised grain size, Equation 4.11(a). The specific application of titanium nitride technology is dealt with in Chapter. 6.

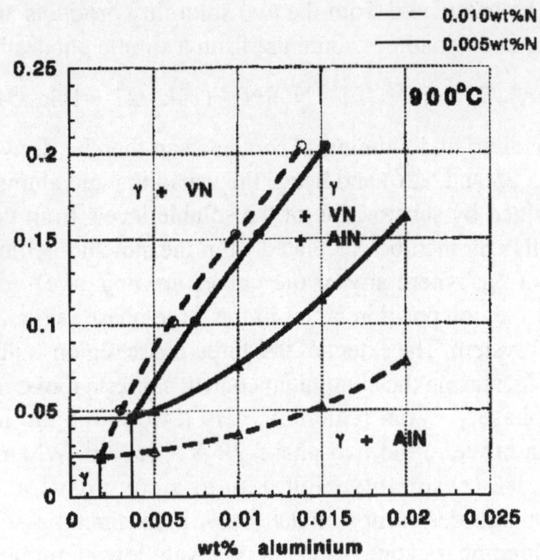

Fig. 5.9 Calculated Phase Fields in the Iron-Rich Corner of the Quaternary Fe-Al-V-N System at 0.005 and 0.010wt%N. The limits of the Three Phase γ + AlN + VN region are shown.

5.2.5 Multiple Additions of Grain Refining Additives – Mutually Exclusive Compounds

Often, alloy additions in steel are made for a variety of reasons, such as controlling the transformation characteristics, or controlling the mechanical properties by forming dispersion hardening particles. Under these circumstances, either more than one type of particle may be formed, or a particle having a complex chemistry is formed. A classic example of the formation of different particle types is found in aluminium – vanadium steels. Both aluminium and vanadium form nitrides, but the AlN has an hexagonal structure, whilst VN has the f.c.c. structure. Both aluminium and vanadium will compete for the available nitrogen on the basis of their solubility products and their respective concentrations. From the solubility products –

$$[Al][N] = k_A \qquad \text{(equ. 5.25)}$$

where k_A is the solubility product for aluminium nitride, and

$$[V][N] = k_V \qquad \text{(equ. 5.26)}$$

where k_V is the solubility product for vanadium nitride.

It is self evident that the dissolved nitrogen contents in Equations 5.25 and 5.26 must the same, provided that both aluminium nitride and vanadium nitride are present. The solubility product for aluminium nitride is more than an order of magnitude smaller than that for vanadium nitride, so that, for a given nitrogen level, aluminium nitride will occur at much lower aluminium contents, than the vanadium content required to form vanadium nitride. Because of the depletion of nitrogen in solution when one of the nitrides is formed, the formation of the second nitride would only occur at a higher level of solute than would otherwise be found. It has been shown (Gladman and McIvor, 1972), that the dissolved nitrogen content can be calculated from the two solubility products and the steel chemistry (aluminium, nitrogen, and vanadium contents) from a simple quadratic equation –

$$[N]^2 + (14Al_T/27 + 14V_T/51 - N_T)[N] - (14k_A/27 + 14k_V/51) = 0 \qquad \text{(equ. 5.27)}$$

The dissolved vanadium and aluminium contents can then be derived from the solubility products, (Equations 5.25 and 5.26) and hence the vanadium and aluminium contents present as nitride can be obtained by substraction of the soluble levels from the totals. The nitrogen present as VN and AlN can then be calculated from the stoichiometric ratios V:N and Al:N. Unrealistic solutions (e.g. where any of the values are negative), give a clear indication (Gladman, 1997) that the composition is not in the three phase ragion, γ + AlN + VN in the quaternary Fe-Al-V-N system. The extent of this three-phase region is illustrated in Fig. 5.9. At low vanadium levels, increasing the aluminium content causes a move from the single-phase γ region into the two-phase γ + AlN region. At very low aluminium levels, the addition of vanadium results in a move to the two-phase γ + VN region. When both aluminium and vanadium contents are reasonably high, e.g. at a composition of 0.15wt.%V with 0.015wt.%Al, then both nitrides occur simultaneously in the three-phase γ + AlN + VN region. The position of this three-phase zone occurs at relatively low aluminium contents, reflecting the higher stability of AlN in comparison with VN. The effect of increasing the nitrogen content of the steel extends the range of the three-phase field to lower aluminium and vanadium contents, as shown in Fig. 5.9. This effect is expected from the solubility products for the individual nitrides, and causes a marked shrinkage of the single-phase (austenite) zone.

The important feature of Equation 5.27, however, is that the soluble nitrogen can be calculated, and from the respective solubility products the soluble vanadium and aluminium contents can be calculated (Equations 5.25 and 5.26). The amounts of vanadium and aluminium present as their respective nitrides are obtained simply by substraction from their total contents. In fact, once the soluble nitrogen content has been determined from Equation 5.27, the procedure for calculating [V] or [Al], and V_{NV} or Al_{AlN}, are exactly the same as described in Section 5.2.3.

5.2.6 Mutually Soluble Components

As mentioned in Section 5.2.4, the carbides and nitrides of the micro-alloying elements, niobium, titanium, and vanadium, all have f.c.c. crystal structures, and all have similar lattice parameters. It is not surprising therefore, that they show complete mutual solid solubility. The complexities of multiple additions of these elements, either with or without an aluminium addition, have been dealt with elsewhere, (Adrian, 1990, 1995). In order to illustrate the concept of dealing with mutually soluble components, we will consider the case of niobium carbo-nitride. All steels contain some carbon and nitrogen, and the addition of the single micro-alloying additive niobium may result in reaction with both carbon and nitrogen to form a niobium carbo-nitride.

Consider the two reactions –

$$[Nb] + [C] \rightarrow (NbC) \qquad \text{(equ. 5.28)}$$

$$[Nb] + [N] \rightarrow (NbN) \qquad \text{(equ. 5.29)}$$

and from these reactions we can write the expressions for the solubility product –

$$k_C = a_{[NB]} . a_{[C]} / a_{(NbC)} \qquad \text{(equ. 5.30)}$$

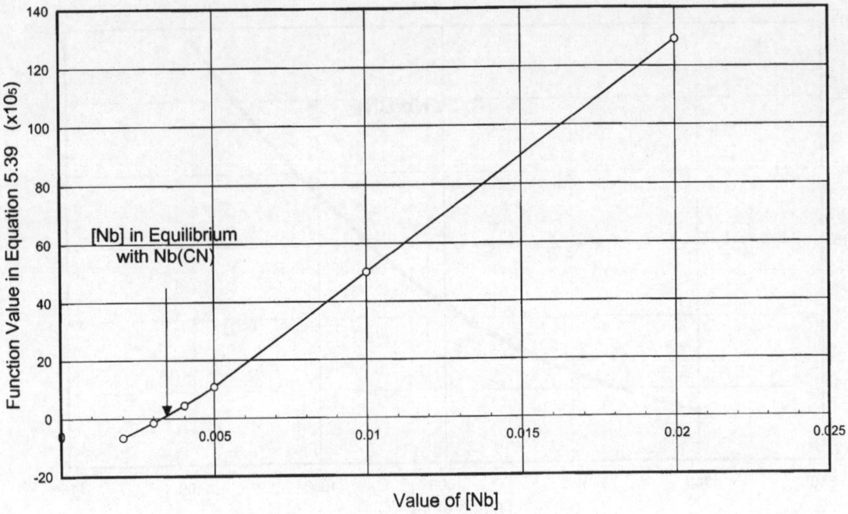

Fig 5.10 Dissolved Niobium at 900°C in Equilibrium with Nb(CN).

$$k_N = a_{[Nb]} \cdot a_{[N]} / a_{(NbC)} \quad \text{(equ. 5.31)}$$

where k_C is the solubility product for niobium carbide,

and k_N is the solubility product for niobium nitride.

However, where the carbides and nitrides are formed, they appear as the carbo-nitride Nb(C,N). The main feature arising from this (Hudd, Jones, and Kale, 1971) is that the activities of the carbide and the nitride in Equations 5.30 and 5.31 are not unity, but depend upon their respective proportions in the carbo-nitride. If we let the proportions of carbide and nitride be x and (1-x) respectively, assuming as before that the activity coefficients for the dissolved carbon, nitrogen, and niobium are unity, and additionally assume that the solid solution of niobium nitride in niobium carbide is ideal, then equating the activities of the carbide and nitride with their respective fraction in the carbo-nitride, we can write the solubility products as –

$$k_C = [Nb][C] / x \quad \text{(equ. 5.32)}$$

$$k_N = [Nb][N] / (1-x) \quad \text{(equ. 5.33)}$$

Now, raising Equation 5.32 to the power x, and Equation 5.33 to the power (1-x), and multiplying the equations gives –

$$k_C{}^x . k_N{}^{(1-x)} = [Nb^x[C]^x[Nb]^{(1-x)}[N]^{(1-x)}/x^x(1-x)^{(1-x)} \quad \text{(equ. 5.34)}$$

or

$$k = [Nb][C]^x[N]^{(1-x)} = k_C{}^x k_N{}^{(1-x)} x^x (1-x)^{(1-x)} \quad \text{(equ. 5.35)}$$

where k is the solubility product of the carbo-nitride.

In logarithmic form –

$$\log k = \log[Nb][C]^x[N]^{(1-x)} = x \log k_C + (1-x) \log k_N + x \log x + (1-x) \log (1-x) \quad \text{(equ. 5.36)}$$

which has the correct thermodynamic form.

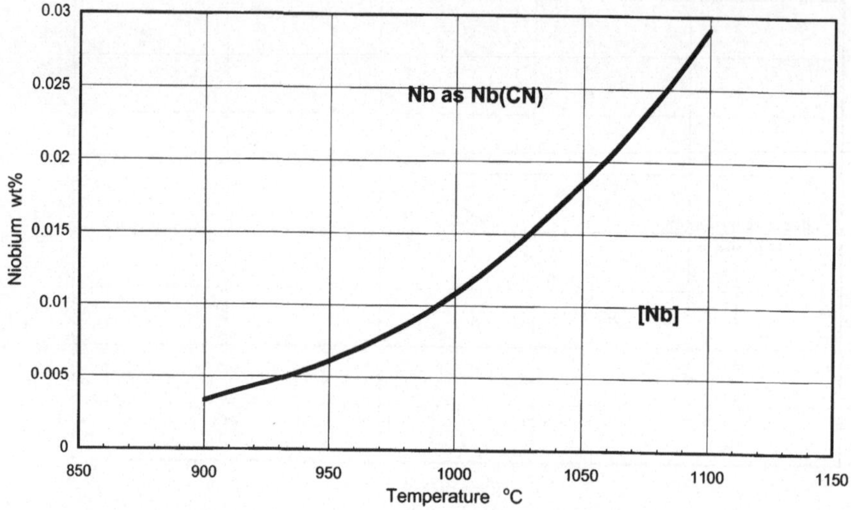

Fig 5.11 Effect of Temperature on the Dissolved Niobium Content of a 0.1wt%C-0.005wt%N-0.03wt%Nb Steel.

Using the mass balance equations and stoichiometric ratios –

$$Nb_T = [Nb] + Nb_{NbCN}$$
$$C_T = [C] + C_{NbCN}$$
$$N_T = [N] + N_{NbCN}$$

$$Nb_{NbC} = x.Nb_{NbCN}$$
$$Nb_{NbN} = (1\text{-}x)\,Nb_{NbCN}$$
$$C_{NbC} = 12.Nb_{NbC}/93$$
$$N_{NbC} = 14.Nb_{NbN}/93$$

Equations 5.32 and 5.33 can be expressed in the form –

$$xk_C = [Nb](C_T - x(Nb_T - [Nb])12/93) \qquad \text{(equ. 5.37)}$$
$$(1\text{-}x)\,k_N = [Nb](N_T - (1\text{-}x)(Nb_T - [Nb])14/93) \qquad \text{(equ. 5.38)}$$

Expanding these equations and equating the terms containing x, with those not containing x, allows expansion of Equation 5.38 to be divided by the expansion of Equation 5.37, thereby eliminating x. This results in a quartic equation –

$$-(12 \times 14)[Nb]^4$$
$$+(2Nb_T \times 12 \times 14 - C_T \times 14 \times 93 - N_T \times 12 \times 93)[Nb]^3$$
$$+(12 \times 93 \times k_N + N_TNb_T \times 12 \times 93 - Nb_T^2 \times 12 \times 14 + C_TNb_T \times 14 \times 93$$
$$+(14 \times 93 \times k_{C)})[Nb]^2$$
$$+(N_Tk_C \times 93 \times 93 - Nb_Tk_C \times 93 \times 14 - Nb_Tk_N \times 93 \times 12 + C_Tk_N \times 93 \times 93)[Nb]$$
$$-(k_Ck_N \times 93 \times 93) = \qquad \text{(equ. 5.39)}$$

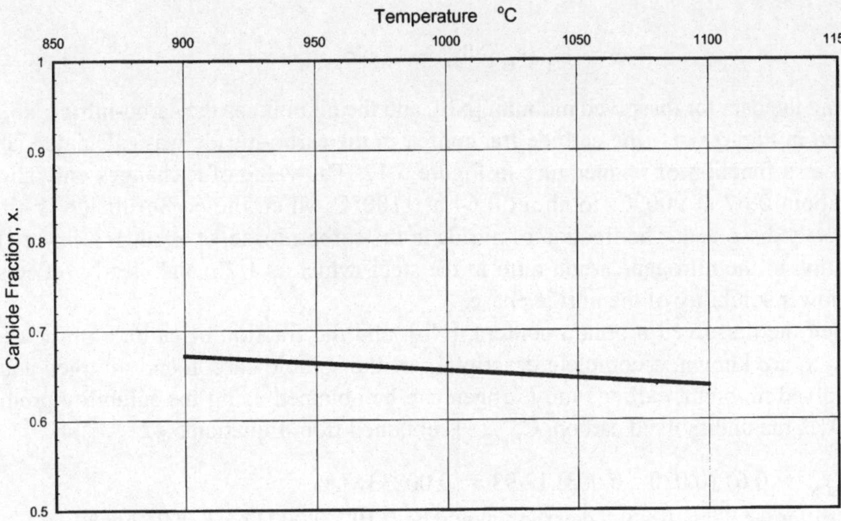

Fig 5.12 Carbide Fraction, x, in Nb(CN).

This can be solved by an iterative procedure, using realistic values of [Nb] between 0 and Nb_T. For example, if we set the left hand side of Equation 5.39 equal to y, then for every assumed value of [Nb], there will be a corresponding value of y. Figure 5.10 shows the values of y, as a function of the assumed values of [Nb], with Nb_T = 0.03wt.%, C_T = 0.10wt.%, and N_T = 0.005wt.%. The temperature selected was 900°C, giving solubility products of k_C = 0.00048 and k_N = 0.000036, in accord with the published solubility data (Narita, 1975) shown in Table 5.2.

This procedure was repeated at various temperatures to show the temperature dependence of the dissolved niobium content in equilibrium with niobium nitride in a steel of this composition (0.1wt.%C – 0.005wt.%N – 0.03wt.%Nb), Figure 5.11. The dissolved niobium content increases from about 0.003wt.% at 900°C, to 0.029wt.% at 1100°C. More importantly, from a grain refinement standpoint, the niobium present as the carbo-nitride, decreases from 0.027wt.% to 0.001wt.% over this same temperature range. It can be seen from Figure 5.11, that the carbo-nitride would be completely dissolved at a temperature of about 1105°C. This decrease in the particulate carbo-nitride would clearly have serious implications for grain size control.

The fraction of carbide, x, in the carbo-nitride can be obtained from the mass balance for carbon –

$$C_T = [C] + C_{NbCN} \qquad \text{(equ. 5.40)}$$

from the solubility product –

$$[C] = xk_C/[Nb] \qquad \text{(equ. 5.41)}$$

and from stoichiometry –

$$C_{NbCN} = x.Nb_{NbCN}.12/93 \qquad \text{(equ. 5.42)}$$

Substituting from Equations 5.41 and 5.42 into Equation 5.40 gives –

$$C_T = x.k_C/[Nb] + x.Nb_{NbCN}.12/13 \qquad \text{(equ. 5.43)}$$

Rearranging Equation 5.43 gives –

$$x = C_T/(k_C/[Nb] + 12.Nb_{NbCn}/93 \quad \text{(equ. 5.44)}$$

Using the data for dissolved niobium [Nb], and the niobium as the carbo-nitride Nb_{NbCN}, depicted in Figure 5.11, the carbide fraction, x, in the carbo-nitride was calculated, and is shown as a function of temperature in Figure 5.12. The value of x changes only slightly from about 0.67 at 900°C, to about 0.64 at 1100°C, when the carbo-nitride is almost completely dissolved. The fraction of nitride in the carbo-nitride, at about 1/3, is very high in relation to the nitrogen/carbon ratio in the steel, which is 1/20, and clearly reflects the much lower solubility of the nitride phase.

When the dissolved niobium content, [Nb], and the fraction of carbide in the carbo-nitride, x, are known, a complete description of the soluble carbon and nitrogen and the undissolved niobium, carbon, and nitrogen can be obtained using the solubility products. At 900°C, the undissolved carbon, C_{NbCN}, is obtained from Equation 5.42.

C_{NbCN} = 0.67.(0.030 – 0.003).12/93 = 0.00233wt.%.

By difference, the dissolved carbon would be 0.10 – 0.00233 = 0.09767wt.%.

Similarly, the undissolved nitrogen can be obtained from –

N_{NbCN} = 0.33.(0.030 – 0.003)14/93 = 0.00134wt.%.

And again, by difference, the dissolved nitrogen would be 0.005 – 0.00134 = 0.00366wt.%. The niobium carbo-nitride content would thus be obtained from the niobium, carbon, and nitrogen figures as 0.027 + 0.00233 + 0.00134 = 0.03067wt.%.

The proportion of niobium carbo-nitride is of course sensitive to the carbon, nitrogen, and niobium contents. Increasing any of these would increase the amount of carbo-nitride or the temperature at which it would dissolve completely.

Bearing in mind the many assumptions that have been made about the activity coefficients for the elements involved, and the absence of any regard for the effects of other elements, such as manganese and silicon, on the activities of nitrogen, carbon and niobium, it seems remarkable that there is very good agreement between the thermo-dynamic predictions described here, and the experimental observations. The solute contents, the amount of precipitate, and the particle chemistry all show good agreement with the predicted values. The

Table 5.3 Densities of Carbides, Nitrides, and Steel, and the Conversion Factors for Wt% to Volume Fraction.

Compound	Density, ρ $Mg\ m^{-3}$	Conversion Factor C. F.
Niobium Carbide	7.84	0.0104
Niobium Nitride	8.41	0.0097
Vanadium Carbide	5.83	0.0140
Vanadium Nitride	6.18	0.0132
Titanium Carbide	4.89	0.0167
Titanium Nitride	5.42	0.0150
Aluminium Nitride	3.27	0.0249
γ- Iron (f.c.c.)	8.15	—

$C.F. = 0.01(\rho_{iron}/\rho_{compound})$

$f_v = C.F.(wt\%)$

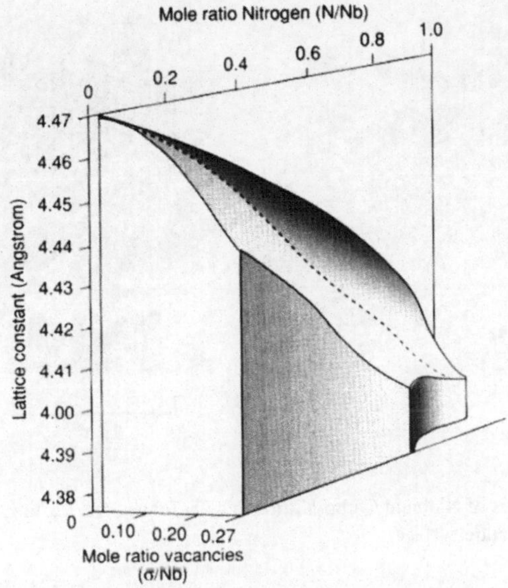

Fig. 5.13 The Effect of Carbon, Nitrogen and Vacancy Concentration, on the lattice Parameter of Niobium Carbo-nitride. The density should reflect this response. (Storms and Krikorian, 1960).

treatment described here for niobium carbo-nitride can be applied, for example, to the niobium-vanadium carbon system, or other similar systems. However, for more complex systems, the reader is directed to the more comprehensive treatments (Adrian, 1990, 1995) which can deal with niobium-vanadium-titanium carbo-nitride in the presence of aluminium nitride, and which also recognises the effects of other elements on the activities of the reactants.

5.3 CONVERSIONS TO VOLUME FRACTIONS

Studies of the solubilities of second phase particles have traditionally developed from a chemico-metallurgical background, whereas studies of grain growth and inhibition have developed from a physics background. The conversion of weight percentage to volume fraction is clearly dependent upon the density of the particulate material relative to the matrix.

Let the weight percentage of NbC be m. Thus for every 100gms. of alloy, there will be m gms. of NbC and (100-m)gms. of iron. Thus the volume of iron, V_{Fe}, and the volume of NbC, V_{NbC}, are –

$$V_{Fe} = (100 - m)/\rho_{Fe}$$

$$V_{NbC} = m/\rho_{NbC}$$

Where ρ_{Fe} is the density of iron

and ρ_{NbC} is the density of NbC

The volume fraction of NbC is defined as the volume of NbC divided by the sum of the volumes of iron and NbC, or –

$$f_V = (m/\rho_{NbC})/\{(100 - m)/\rho_{fe} + m/\rho_{NbC}\} \quad \text{(equ. 5.45)}$$

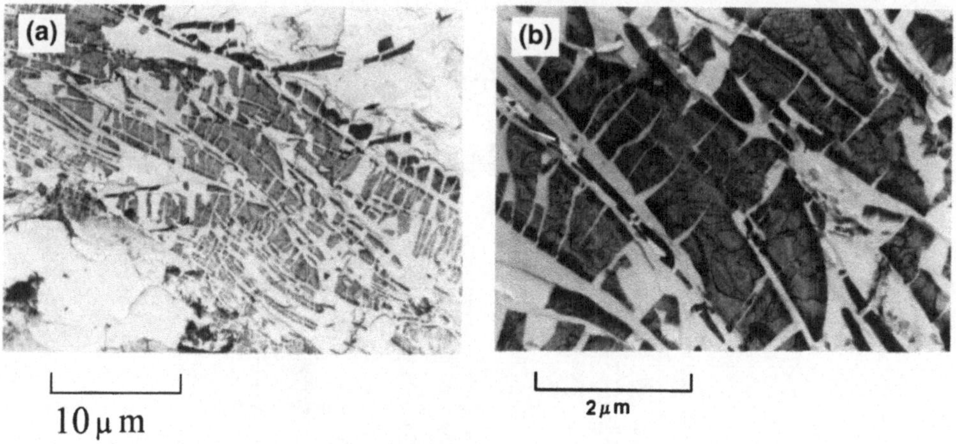

Fig. 5.14 Extensive Flakes of Niobium Carbo-Nitride in a 25-Tonne Steel Casting. Extraction replica taken from an inter-granular fracture surface.

If m is very small, e.g. less than 0.1wt.%, then the relationship can be approximated by –

$$f_V = m\rho_{Fe}/100\rho_{NbC} \qquad \text{(equ. 5.46)}$$

The densities of the various carbides and nitrides and iron are shown in Table 5.3. The density of the γ-phase is shown rather than that of the α-phase, as much of the grain control is effected in the high temperature region where the f.c.c. γ-phase (austenite) is stable. If grain size control is applied in the b.c.c. α-iron (ferrite) region, the density is 7.85 Mgm^{-3}, i.e. about 3.7% less than that of austenite. If should be noted that the room temperature densities of the various phases are quoted and these may change, albeit very slightly, at the relevant temperatures. The densities of the commonly used carbides and nitrides vary from 3.27 to 8.41 Mgm^{-3} thus showing a variation in volume fraction of $2^1/_2$ times for a given weight percentage of a particular phase.

From Equation 5.46, it is obvious that the weight percentage m can be converted to the volume fraction by using a conversion factor, CF, where –

$$CF = 0.01\rho_{Fe}/\rho \text{ compound} \qquad \text{(equ. 5.47)}$$

So that -

$$f_V = CF \times (\text{wt.\% of compound}) \qquad \text{(equ. 5.48)}$$

The conversion factors for the various compounds are also shown in Table 5.3. In the event of a complex compound being formed, e.g. Nb(C,N), a linear relationship between the density and the carbide fraction is often assumed, the terminal densities being those of the carbide and the nitride given in Table. 5.3. This is not necessarily correct (Storms and Krikorian, 1960), and the situation can be further complicated by relatively high vacancy concentrations as shown in figure 5.13. Although changes in lattice parameter are shown in Figure 5.13, these are directly linked to density changes, provided that the vacancy concentration remains low. Nevertheless, the linear approximation, neglecting vacancy concentrations, does not appear to cause any serious errors as far as grain size control is concerned.

5.4 PARTICLE SIZE AND PARTICLE COARSENING

The earlier sections of this chapter have dealt with the chemical and thermodynamic factors which control the volume fractions of particles that are so important in grain size control as illustrated by Equation 4.11(a) – (c). The second important feature of the particles that is equally important to grain size control is the particle size. There are two important stages which define the particle size, the first being the precipitation of the particles from solid solution, and the second being the coarsening of particles after precipitation has been completed. The latter process is known as 'Ostwald Ripening'.

5.4.1 PRECIPITATION

Precipitation of second phase particles can take place in many ways, some of which could have the direst consequences for grain size control. The most important feature affecting the precipitation process is super-saturation. A high level of super-saturation will activate a large range of nucleation sites. The general range of nucleation sites will include grain boundary nucleation, dislocation nucleation, and homogeneous nucleation, the nucleation difficulty increasing in the order given. The classic example of homogeneous nucleation is in the duralumin type alloys, (Guinier, 1938, Preston, 1938). The formation of the θ-phase ($CuAl_2$), in the matrix has been followed in considerable detail. An account of other systems showing homogeneous precipitation has been presented (Sinha, 1989). A common feature of these materials is the use of a solution treatment followed by ageing at a lower temperature, to allow precipitation of the solute from a highly super-saturated solution. Although similar treatments can be applied to microalloyed steels, the situation is somewhat different due to the allotropic change from the f.c.c. austenite at the solution temperature to the b.c.c. ferrite at typical precipitation temperatures (below 700°C).

In the case of the microalloyed carbo-nitrides in steel, there are several precipitation processes. During extremely slow cooling, precipitation occurs at low levels of super-saturation, and the carbo-nitride is precipitated as extensive plates in the grain boundary plane, Figure 5.14. This precipitate morphology was found in a 25 ton sand-cast node, and these particles cannot be used effectively for grain size control, because the plates only occur on the grain boundaries of the very coarse (several hundred microns), as-cast austenite grain structure.

At somewhat faster cooling rates, the precipitation of these carbo-nitrides (and aluminium nitride) in austenite, is extremely sluggish, despite the increased super-saturation, but deformation of the super-saturated austenite accelerates the precipitation process as shown in Table 5.4, (Simoneau, Begin, and Marquis, 1978). The introduction of

Table 5.4 Precipitation of Nb(C,N)in Austenite at 900°C in a 0.04wt%Nb Steel.

Hot Deformation*	Time for 50% Precipitation (seconds)	Reference
0% – Held isothermally	16,800	Hoogendorn and Spanraft, 1977
Rolled 17% and held	678	Simoneau et al, 1978
Rolled 33% and held	174	—
Rolled 50% and held	38	—
Rolled 67% and held	19	—

* *Soaked at 1250°C prior to hot rolling.*

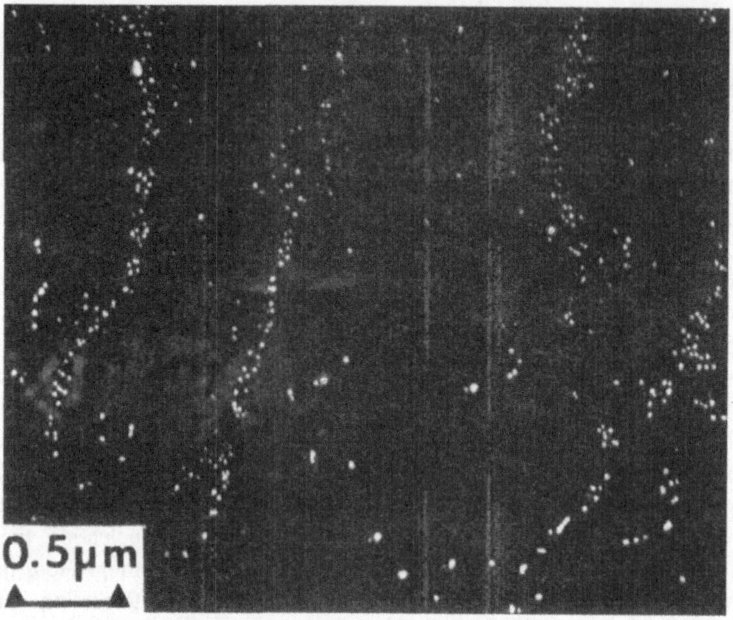

Fig. 5.15 Fine Niobium Carbo-Nitride Particles Precipitated During the Hot-Working of Steel. The particles decorate the sub-grain boundaries of the prior austenite. Dark field transmission electron micrograph centred on the (111) Nb(C,N) reflection.

a relatively small deformation of 17% drastically reduces the time for 50% precipitation from over 4 hours to about 11 minutes. Increasing the deformation further to 67% reduces the time for 50% precipitation to only 19 seconds. The reason for the rapid acceleration of precipitation is readily explained by the observation of niobium carbo-nitride particles precipitated on the low angle sub-grain boundaries that result from recovery processes in the hot deformed austenite. Figure 5.15 (Santella, 1981) shows fine niobium carbo-nitrides decorating the sub-grain boundaries. The sub-grains appear to be elongated and are 0.5 to 1.0μm in width. The carbo-nitride particles themselves appear to be of the order of 10 – 20nm in diameter.

If the niobium carbo-nitride is dissolved and cooled through the austenite temperature range at a rate which precludes substantial precipitation, the carbo-nitrides are often precipitated on the austenite/ferrite interface during the allotropic transformation, giving the characteristic interphase particle distribution, (Pickering and Gladman, 1963), as shown in Figure 5.16. The particles can be seen as rows, which are really sheets of particles seen 'edge on' in the foil. The rows are about 25nm apart, and the particles are of the order of 2.5nm in diameter. Although the spacing of particles within the rows appears to be very small, it must be remembered that they are sheets of particles viewed edge on. A more realistic impression of the particle spacing is given by the bulging portions of the dislocation seen in Figure 5.16, each bulge being more than 0.1μm apart, although even here, the particle spacing is foreshortened because the dislocation is at an angle to the plane of the foil.

The particle size will obviously be affected by cooling rates, holding time, and steel chemistry, but it is not the objective to give a complete account of the precipitation

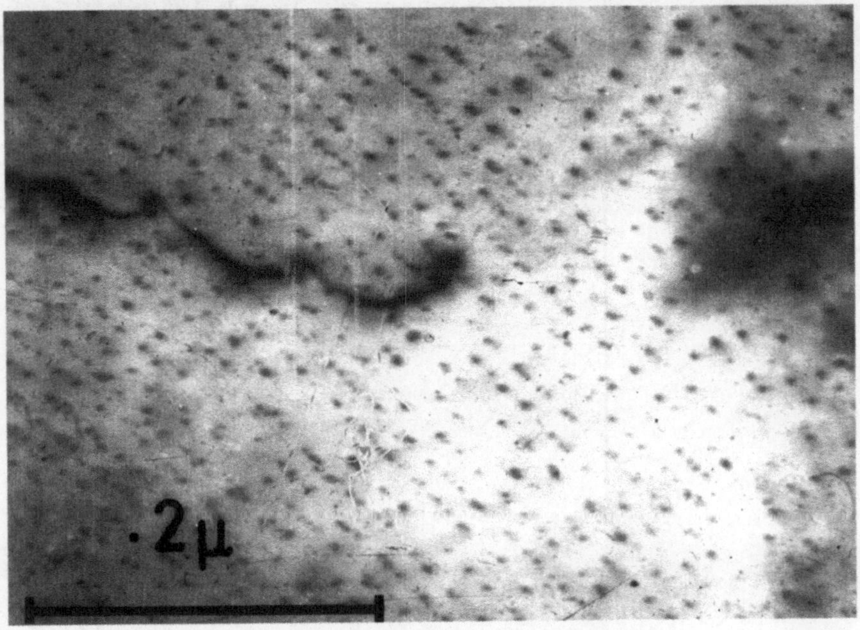

Fig. 5.16 Interphase Precipitation of Niobium Carbo-Nitride During Transformation of Austenite to Ferrite on Cooling. Note the interaction with the dislocation; this gives some indication of the real spacing of the particles.

processes here. Rather, the objective is to illustrate that considerable variations in particle size may occur as a result of the particulars of the precipitation process, and that the success of any grain size control attempt may be dependent upon these particulars. For example, the coarse dispersion of Nb(C.N) in the large casting referred to earlier could not be used to inhibit the growth of grains a few microns in diameter, whereas the fine dispersions of particles induced by hot working, or by interphase precipitation, would form a very suitable starting point for grain size control, as will be shown in Chapter 6.

5.4.2 Ostwald Ripening

Particle size is one of the factors defining the boundary between grain growth/inhibition. Any increase in particle size at elevated temperatures may allow grain growth to occur in what had been a stable granular structure. Ostwald ripening is just such a process in which the average particle size increases. The definition of Ostwald ripening is the selective growth of the larger particles in a particulate array, accompanied by the shrinkage and ultimate disappearance of the smaller particles, the total volume fraction of particles remaining substantively unchanged. This process, like grain growth, reduces the interfacial area between the particles and matrix, and thereby causes a significant reduction of energy. The difference between the two processes is that grain growth is a tactile process in that each grain is in contact with its neighbours and is aware of its relative size by the nature of the contact angles or boundary curvature. This is not the case in particle coarsening, where each particle is separated from neighbouring particles by a not inconsiderable thickness of the matrix phase. This poses the question of how any particular particle knows whether there is

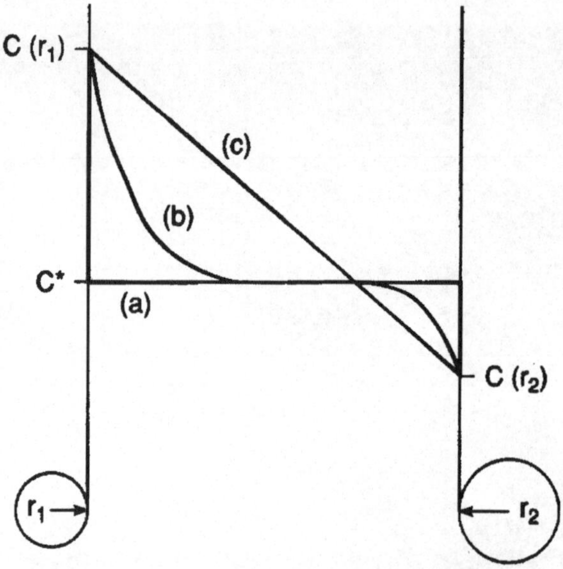

Fig. 5.17 Ostwald Ripening. The solute profiles between particles of different sizes are shown for different rate controlling factors – (a). Interface control. (b). Diffusion control with large spacing. (c). Diffusion control with small spacing.

a larger particle nearby, in which case it should shed atoms and start to dissolve, or whether there is a smaller particle nearby in which case it should start to collect atoms and grow.

The answer to this question is to be found in the Gibbs-Thomson, or Thomson-Freundlich equation, (Freundlich, 1922) which describes the local equilibrium between the solute content and the particle size at the matrix particle interface.

$$\ln(S_r/S) = 2\Omega\sigma/RTr \qquad \text{(equ. 5.49)}$$

where S_r is the solute content in equilibrium with a particle of radius, r.
S is the true equilibrium solute content with a particle of infinite size
Ω is the molar volume of the precipitate
σ is the interfacial energy
R is the gas constant
and T is the absolute temperature, K.

The derivation of this equation has been given elsewhere (Martin and Doherty, 1976). The natural logarithm of the equilibrium solute content varies inversely as the particle radius, so that large particles have a lower equilibrium solute content than smaller particles. The solute concentration will vary through the matrix, and the nature of the variation will depend upon the rate controlling diffusion steps, as indicated in Figure 5..17, (Fischmeister and Grimvall, 1973). This shows the solute profiles that are expected when interface controlled diffusion, bulk diffusion with small inter-particle spacings, or bulk diffusion with large inter-particle spacing, are operative. The transport of solute from small particles to large ones may also occur by grain boundary diffusion, or by pipe diffusion along dislocations. The common feature, however, is that diffusion of solute away from the high solute concentration regions near small particles, will lower the solute content locally, and the small particle will have to restore the solute to equilibrium level. In so doing, the particle becomes smaller and requires an even higher equilibrium solute content. Conversely, diffusion of solute from the remote matrix into the low solute content regions near large particles will result in precipitation on the large particles to restore equilibrium, and, because of this, the particle will grow and require an even lower equilibrium solute content. These processes will continue, resulting in the disappearance of the small particles and growth of the larger particles.

The plateau of solute concentration in the matrix that is unaffected by the perturbations near the particles, marked C* in Figure 5.17, is intermediate between the equilibrium solubilities for the largest and the smallest particles, and may be equated with the solubility in equilibrium with the average size of particle.

In the case of bulk diffusion controlled coarsening the rate of change of particle size, dr/dt, can be expressed (Wagner, 1961) as –

$$dr/dt = D\Omega\, ds/dx \qquad \text{(equ. 5.50)}$$

where D is the diffusion coefficient of the solute

Ω is the molar volume

and ds/dx is the concentration gradient near to the particle surface.

Now the volume fraction of particles remains substantially constant, i.e. –

$$4\pi\Sigma r_i^2\,(dr_i/dt) = 0 \qquad \text{(equ. 5.51)}$$

where r_i is the particle radius of the "i" th class.

From Equation 5.50 and using an approximation for the stationary diffusion flux from an infinite medium of concentration, S*, to that, S_r, at a particle of radius r, yields (Zener, 1949) –

$$dr/dt = D\Omega\,(S^* - S_r)/r \qquad \text{(equ. 5.52)}$$

Substitution from Equations 5.49 and 5.51 gives (Fischmeister and Grimvall, 1973) –

$$dr_i/dt = \sigma\, S\, D\Omega^2\,(1/r^* - 1/r_i)/RTr \qquad \text{(equ. 5.53)}$$

where σ is the interfacial energy

r* is the critical particle radius

S is the solute content in equilibrium with the critical particle radius.

The critical particle radius can be equated with the average particle radius. Despite the simplicity of the concept, and the simple portrayal of the solute profile in Figure 5.17, the mathematical description of the change in mean particle size with time is quite difficult (Lifshitz and Slyozov, 1961, Wagner, 1961, Greenwood, 1956). The Wagner equation is given as –

$$r^3 - r_0^3 = 8D\sigma\Omega St/9RT \qquad \text{(equ. 5.54)}$$

where r is the mean particle radius after a time t

r_0 is the initial mean particle radius

D is the diffusion coefficient

σ is the interfacial energy

Ω is the molar volume
S is the equilibrium solute content
(in equilibrium with a particle of mean radius) expressed as the atom fraction
R is the gas constant
and T is the absolute temperature, K.

This equation is only valid for relatively widely dispersed particles, but the volume fractions and particle sizes that are involved in grain size control do not invalidate the use of the Wagner equation. The use of this equation in a relatively simple system such as the Cu-Cr binary, where particles of chromium are precipitated in the copper matrix, is fairly clear. The solute level would refer to the chromium content of the copper matrix, and the diffusion coefficient is that of Cr in Cu. The precipitation of Ni_3Al or NiAl compounds in a nickel matrix would also appear to be relatively straightforward, the solute being aluminium although the molar volume would relate to the compound. In the case of micro-alloyed steel, where particles such as NbC exist in an iron matrix, there are two solutes, i.e. niobium and carbon. This matter has been considered (Fishmeister and Grimvald, 1973) who pointed out that the Thomson-Freundlich equation is modified by inserting the solubility product rather than the elemental solubility, i.e. –

$$\ln (k_r/k_s) = 2\Omega\sigma/RTr \qquad \text{(equ. 5.55)}$$

where k_r is the solubility product in equilibrium with a particle of radius, r.

However, under conditions of diffusion limited growth , particle coarsening may be limited by the component with the lowest diffusion flux in relation to its concentration in the precipitated phase.

Whilst bulk diffusion control of particle growth can be expected in many systems, it is important to acknowledge that other rate controlling mechanisms exist. For example, the transfer of atoms across the interface may be rate limiting, and interface 'poisons' can inhibit the transfer of solutes. An example of this type of effect is to be found when silicon concentrates at the interface between iron and iron carbides in steel. This slows down the coarsening of carbides quite dramatically. When this type of control is operative, the coarsening rate can be described (Wagner, 1961) by the equation –

$$r^2 - r_0^2 = 64\, K_T s\, \sigma\, \Omega^2 t / 81RT \qquad \text{(equ. 5.56)}$$

where K_T is the temperature dependent transfer coefficient.

Other power law equations describe the particle coarsening rates associated with other rate controlling processes. For example, grain boundary diffusion control, which might be expected if all particles occur on grain boundaries, results in an equation (Kirchner, 1971) of the form –

$$r^4 - r_0^4 = kt \qquad \text{(equ. 5.57)}$$

Other studies, (Kreye, 1970) have indicated an r^5 dependence on time for the coarsening of particles on sub-grain boundaries where pipe diffusion along dislocations may be expected to play an important role.

There are other complications that may arise when the precipitation or dissolution processes cause a significant volume change. The classic example of this is found in the coarsening of cementite (Fe_3C) particles in α-iron during the tempering of martensite. It has been suggested (Honeycombe, 1981) that the local volume increase due to the precipitation of Fe_3C, and the local volume decrease due to dissolution of Fe_3C, can only

Table 5.5 Diffusion Data for Micro-alloy and Interstitial Elements, and for the Self-Diffusion of Iron, in Austenite and Ferrite. (*Diffusion coefficients at selected temperatures are also shown.*)

Phase/Element m^2s^{-1}	Do kJ/mol	Q m^2s^{-1}	D(700°C) m^2s^{-1}	D(900°C) m^2s^{-1}	D(1200°C)
Ferrite					
C	$0.62x10^{-6}$	80.4	$3.09x10^{-11}$	$1.63x10^{-10}$	—
N	$0.50x10^{-6}$	77.0	$3.67x10^{-11}$	$1.86x10^{-10}$	—
Ti	$3.15x10^{-4}$	248.0	$1.53x10^{-17}$	$2.85x10^{-15}$	—
V	$0.61x10^{-4}$	267.1	$2.79x10^{-19}$	$7.78x10^{-17}$	—
Nb	—	—	—	—	—
Al	$0.10x10^{-4}$	221.8	$1.24x10^{-17}$	$1.33x10^{-15}$	—
Fe	$1.67x10^{-4}$	256.7	$2.76x10^{-18}$	$6.18x10^{-16}$	—
Austenite					
C	$0.10x10^{-4}$	135.7	—	$9.06x10^{-12}$	$1.59x10^{-10}$
N	$0.91x10^{-4}$	168.6	—	$2.82x10^{-12}$	$0.96x10^{-10}$
Ti	$0.15x10^{-4}$	251.2	—	$9.76x10^{-17}$	$1.86x10^{-14}$
V	$0.25x10^{-4}$	264.0	—	$4.29x10^{-17}$	$1.07x10^{-14}$
Nb	$5.30x10^{-2}$	344.4	—	$2.39x10^{-17}$	$3.19x10^{-14}$
Al	—	—	—	—	—
Fe	$0.49x10^{-4}$	284.1	—	$1.09x10^{-17}$	$4.13x10^{-15}$

be accommodated by a significant vacancy flux from the dissolution sites to the precipitation sites. As a consequence of this, even though the correct r^3 dependence on time is observed, the activation energy observed is that for self diffusion in α-iron rather than the much lower activation energy for the diffusion of carbon in α-iron.

5.4.3 Implications of the Wagner Equation for Particle Selection

The Wagner expression for bulk diffusion controlled growth, Equation 5.54, has important implications for the selection of grain refining agents.

The Diffusion Coefficient, D

Particles containing elements with low diffusion coefficients would give low particle coarsening rates. The Diffusion coefficient, D, is related to temperature by the Arrhenius relationship –

$$D = D_0 \exp(-Q/RT) \qquad \text{(equ. 5.58)}$$

The diffusion data D_0 and Q, for micro-alloying elements and interstitial elements in both f.c.c (austenite) and b.c.c (ferrite iron are shown in Table 5.5, together with diffusion coefficients at selected temperatures in the austenite and ferrite temperature ranges (Smithells, 1983). In the ferrite (α-phase), the diffusion coefficients of the interstitials at 700°C are of the order of 3 x $10^{-11}m^2s^{-1}$, whilst those of the substitutional micro-alloying elements are of the order of $10^{-18}m^2s^{-1}$. In the austenite (γ-phase) the diffusion coefficients of the interstitials at 1200°C are about $10^{-10}m^2s^{-1}$, whilst those of the substitutional solutes are of the order of $10^{-14}m^2s^{-1}$, and are similar to the self diffusion of iron. In micro-alloyed steels, therefore, the diffusion flux of the substitutional micro-alloying elements would be expected to limit the rate of particle coarsening, although the complexities of the variation in solubility product of complex micro-alloy carbo-nitrides with particle size have not yet been exploited.

Table 5.6 Molar Volumes of Carbides and Nitrides, and of Iron.

Compound (Structure)	Molecular Mass	Lattice Parameter (nm)	Molecules Per Unit Cell	Density Mgm^{-3}	Molar Volume (cm^3)
NbC (fcc)	105	0.4462	4	7.84	13.39
NbN (fcc)	107	0.4387	4	8.41	12.72
VC (fcc)	63	0.4154	4	5.83	10.81
VN (fcc)	65	0.4118	4	6.18	10.52
TiC (fcc)	60	0.4313	4	4.89	12.27
TiN (fcc)	62	0.4233	4	5.42	11.44
AlN (cph)	41	c= 0.4965*	6	3.27	12.54
γ- iron (fcc)	56	0.357	4	8.15	6.85
α- iron (bcc)	56	0.286	2	7.85	7.11

* *a= 0.311*

THE MOLAR VOLUME, Ω

The density, ρ, of a micro-alloy carbide or nitride can be obtained from the molecular mass, m_{MX}, the lattice parameter,a, and the number of molecules per unit cell, n.

$$\rho = n.m_{MX} / Aa^3 \qquad \text{(equ. 5.59)}$$

where A is Avogadro's number (6.0222 x 10^{23}). The molar volume, Ω, is obtained from –

$$\Omega = m_{MX}/\rho = Aa^3/n \qquad \text{(equ. 5.60)}$$

The lattice parameters, molecular masses and densities are shown in Table 5.6 for iron, the micro-alloy carbides and nitrides, and aluminium nitride. A striking feature of Table 5.6 is that the molar volumes of the micro-alloy carbides and nitrides all lie within the relatively narrow range of 10.52cm^3 to 13.39cm^3 despite the range of particle densities. This is clearly due to the similarities of their lattice parameters. It is also interesting to note that the molar volume of aluminium nitride, at 12.54cm^3, is well within this range despite the different lattice structure. There is little opportunity to use molar volumes to minimise particle coarsening from these particular compounds as they are all similar.

THE SOLUTE LEVEL, S, OF THE RATE LIMITING SPECIE.

The solute levels of the various species are dependent upon the temperature and alloy composition as described earlier in this Chapter, see Section 5.2. Increasing the holding temperature would not only decrease the volume fraction of particles, but would increase the solute content, S, and thus increase the rate of particle coarsening. Increasing the temperature would also accelerate the diffusion processes, which would also give a pronounced increase in the rate of coarsening.

If we consider the case of aluminium nitride, both the dissolved aluminium and nitrogen contents will increase with increasing temperature, as indicated by the solubility product–

$$[Al]\,[N] = k_S \qquad \text{(equ. 5.61)}$$

$$k_S = k \exp(-Q_S / RT) \qquad \text{(equ. 5.62)}$$

Table 5.7 Solubility and Diffusivity of Vanadium and Nitrogen in Austenite in a 0.15wt%V-0.015wt%N Steel.

(a) Vanadium

Temperature °C	[V] wt%	S Atom Fraction	D m^2s^{-1}	SD m^2s^{-1}
900	0.103	1.13×10^{-3}	4.29×10^{-17}	4.85×10^{-20}
950	0.109	1.20×10^{-3}	13.24×10^{-17}	15.89×10^{-20}
1000	0.118	1.30×10^{-3}	36.72×10^{-17}	47.74×10^{-20}

(b) Nitrogen

Temperature °C	[N] wt%	S Atom Fraction	D m^2s^{-1}	SD m^2s^{-1}
900	0.0021	0.84×10^{-4}	2.82×10^{-12}	2.37×10^{-16}
950	0.0037	1.48×10^{-4}	5.70×10^{-12}	8.44×10^{-16}
1000	0.0062	2.48×10^{-4}	11.10×10^{-12}	27.53×10^{-16}

The greatly increased diffusion rate of the interstitial atoms (Table 5.5) compared with those of the substitutional solutes, would mean that the latter should be rate controlling. However, as indicated in Figure 5.2, the dissolved aluminium content can be changed at any given temperature by alloy composition changes. For example, increasing the nitrogen content of the alloy from X to S and then Y, at a fixed aluminium level, would reduce the dissolved aluminium content, [Al], at 1000°C, as indicated by the points X, S_1, K_1, i.e. the value of [Al] would be reduced from 0.017wt.% to 0.008wt.% when the total nitrogen level is increased from 0.007wt.% to 0.015wt.%. Conversely, if the aluminium content of the alloy was increased, the dissolved nitrogen, [N], at 1000°C would decrease and therefore the dissolved aluminium, [Al], would increase, thereby increasing the rate of coarsening of the aluminium nitride particles.

If the bulk diffusion controlled particle coarsening mechanism is operative, then according to the Wagner equation, (Equation 5.54), the rate of coarsening would be dependent upon the product of the diffusion coefficient D and the solute content, S, of the rate limiting specie. The products D_iS_i for the different solutes would therefore give a clear indication of the coarsening rate, the lowest product being rate limiting. The more complicated situation arising from the presence of complex compounds, e.g. niobium-vanadium carbide, or carbo-nitride, has been dealt with elsewhere (Gladman, 1997).

Interfacial Energies, s

Many of the interfacial energies for incoherent particles measured in ferrous and other metallic systems lie in the range 0.5 – 1.2Jm^{-2}. In the case of coherent particles, the values are in the range 0.1 – 0.2Jm^{-2}. Whilst coherent particles are seen to form during inter-phase precipitation, particularly in vanadium steels, where the lattice mismatch between ferrite and the carbo-nitride is small, the coherency does not appear to survive thermal cycling through the phase transformations, and micro-alloy carbo-nitrides precipitated in austenite will be incoherent after transformation to ferrite, and vice-versa. These particles would, therefore, most probably have interfacial energies in the region of 0.5 – 1.0Jm^2. The small effect of temperature on these interfacial energies can usually be ignored (Fischmeister and Grimvall, 1973).

5.4.4 Calculation of Particle Growth Rates

Having much of the data required by the Wagner equation for bulk diffusion controlled particle growth (Equation 5.54), the particle coarsening rates can be calculated for those cases where bulk diffusion control is assumed to be appropriate. Consider the case of vanadium nitride particles held at various temperatures in the austenite range, in a 0.15wt.%V – 0.015wt.%N steel. The solubility product for VN in austenite (see Table 5.2) is given as –

$$\log k_s = 3.02 - 7840/T,$$

yielding values of $k_s = 2.17 \times 10^{-4}$, 4.07×10^{-4}, and 7.27×10^{-4} for temperatures of 900°C, 950°C, and 1000°C, respectively. Solving the quadratic equations yields the undissolved vanadium contents of 0.0470, 0.0410, and 0.0322wt.% respectively for these same temperatures, which, by subtraction from the total vanadium content, gives corresponding dissolved vanadium contents of 0.1030, 0.1090, and 0.1178wt.%. These are converted to atom fractions, S, by the (reasonable) approximation –

$$S = 0.01(\text{wt.\%V}) \times 56/51$$

where the relative atomic masses of iron and vanadium are 56 and 51 respectively.

The soluble vanadium, S, in the Wagner equation (expressed as atom fractions), at the various temperatures are shown in Table 5.7(a). The diffusion coefficients at the corresponding temperatures are also shown. Given that the interfacial energies and the molar volumes are relatively insensitive to temperature (or even to the type of particle used), then the product of the soluble vanadium and the diffusion coefficient is important in controlling the rate of particle coarsening, provided that coarsening is controlled by vanadium diffusion. The other possibility is that the process is controlled by nitrogen diffusion. Following the same procedure as was used for vanadium, the data for nitrogen is shown in Table 5.7(b). Despite the fact that the solute content for nitrogen is an order of magnitude less than that for vanadium, the diffusion coefficients differ by a factor of 10^5 or more, nitrogen having the higher diffusivity. The rate of growth on a volume basis must clearly be dependent upon the vanadium flux, and is some 10^4 times less than that which would be achieved by the nitrogen flux.

The importance of the solute level in the Wagner equation is illustrated by comparing the calculated coarsening rates of titanium and vanadium nitrides, see Figure 6.11, where the very low solubility of titanium nitride produces much lower coarsening rates (Gladman, 1997).

REFERENCES

Adrian H., 1990. Microalloyed Vanadium Steels, eds. M. Korchynsky, S. Gorczca, and M Blichararski, Akad. Gom. – Hutz, Krakow, Poland, p.105.

Adrian H., 1995. Microalloying '95, I.S.S. Pittsburgh, p.285.

Fischmeister H. and Grimvall G., 1973. Materials Science Res. Vol.6, p.119.

Freundlich H., 1922. Kapillarchemie, Akad. Verlagsges, Leipzig.

Gladman T., 1997. The Physical Metallurgy of Microalloyed Steels, Inst. of Mat. London.

Gladman T. and McIvor I.D., 1972. Scandinavian Journal of Metallurgy, vol.1, p.247.

Greenwood G.W., 1956. Acta Met, vol.4, p.243.
Guinier A., 1938. Nature (London) vol.142, p.142.
Honeycombe R.W.K., 1981. Steels – Microstructure on Properties, Edward Arnold, London.
Hoogendorn T.M., and Spanraft M.H., 1977. Microalloying '75, Ed. M. Korchynsky, Union Carbide Corp., New York, p.75.
Hudd R., Jones A., and Kale M.N., 1971. Journ. Iron Steel Inst., vol.209, p.121.
Irvine K.J., Pickering F.B. and Gladman T., 1967. Journ Iron Steel Inst., vol.205, p.161.
Kreye H., 1970. Z. fur Metallkunde, vol. 61, p.108.
Leslie W.C., Ricketts R.L, Dotson C.L., and Walton C.S., 1954. Trans A.S.M, vol.46, p.1470.
Lifshitz I.M., and Slyozov V.V., 1961. Journ. Phys. Chem. Solids, vol.19, p.35.
Martin J. W. and Doherty R.D., 1976. Stability of Microstructure in Metallic Systems, Cambridge University Press, Cambridge.
Narita K., 1975. Trans Iron and Steel Inst. of Japan, vol.15, p.145.
Pickering F.B. and Gladman T., 1963. Iron and Steel Inst., Special Report No. 81, London, p.10.
Preston G.D., 1938. Proc. Roy. Soc., London, vol. A167, p.526.
Santella M.L., 1981. Ph.D. Thesis, Dept. of Met. And Mat. Eng. University of Pittsburg.
Simoneau R., Begin G., and Marquis A.H., 1978. Met. Sci., vol.12, p.381.
Sinha A.K., 1989. Ferrous Physical Metallurgy, Butterworths, Stoneham, Mass.
Smithells C.,1983. Smithell's Metals Reference Book (Sixth Edition), Ed. E. A. Brandes, Butterworths, London.
Storms E. K. and Krikorian N. H., 1960. J. Phys. Chem., vol.64, p.1747.
Wagner C., 1961. Z. Electrochemie, vol.65, p.243.
Zener C., 1949. Journ. App. Phys., vol.20, p.950.

Chapter 5 Exercises

1. 0.1wt.%V is added to a 0.01wt.%N nitrogen steel, which is then heated to 900°C and equilibrated at this temperature. (a) What is the equilibrium constant at this temperature? (b) What is the percentage by weight of vanadium present as vanadium nitride? (c) What is the percentage by weight of vanadium nitride? (d) What is the volume fraction of vanadium nitride? Assume that the carbon content can be neglected.

2. A steel containing 0.03wt.%Nb – 0.1wt.%C – 0.005wt.%N is held at 1000°C until equilibrium is attained. What is the wt.% of niobium as the carbo-nitride ? What is the carbon/nitrogen ratio in the carbo-nitride (by weight)?

3. A 0.01wt.%Ti – 0.005wt.%N steel is held at 1250°C for 3 hours prior to hot rolling (a) What is the dissolved titanium content (wt.%)? (b) What size would you expect the TiN particles to be after the 3 hour hold, given that the initial radius was 10nm?

CHAPTER 6

Applications in Grain Size Control

The principles of grain growth and the various approaches to this phenomenon were described in Chapter 3, whilst the inhibition of grain growth, both in bulk material and in foils was described in Chapter 4. Particular attention was directed to the inhibition effects of second phase particles, whilst Chapter 5 was concerned with the precipitation behaviour of second phase particles, and with subsequent coarsening of the particles on holding at elevated temperature. Although much of the detailed material concerned the grain refinement of steel, reflecting the author's particular area of study, the principles are applicable to any polycrystalline material. In this, the final chapter, the application of these principles are considered in relation to grain growth following allotropic transformations and grain growth following hot work, or cold work and re-crystallisation. The effects of particles on the re-crystallisation process itself is also considered in terms of the inhibition of sub-grain growth.

6.1 STEEL NORMALISING TREATMENTS

The allotropic change from the b.c.c. ferrite to the f.c.c. austenite on heating a plain carbon steel causes a complete re-crystallisation of the structure, as does the reverse transformation from austenite to ferrite during cooling to ambient temperature. Variations in microstructure caused by variations during the hot working processes used in shaping can be 'normalised' by heating the stock material to the normalising temperature in the austenite range, holding at this temperature, and then air cooling, as described in Chapter 4. The main feature of a normalising treatment, however, is to produce a consistent and reproducible microstructure, which in turn produces consistent and reproducible material properties. The epic work of Hall and Petch (Hall, 1951) (Petch, 1953) directed attention to the important effects of grain size on the strength and toughness of structural steels. Having a well tried thermal treatment such as normalising in regular usage, efforts to improve properties were initially directed at normalised materials.

The allotropic change in steels occurs by a nucleation and growth transformation, the favoured nucleation sites being carbide particles and ferrite grain boundaries during heating to form austenite, and austenite grain boundaries during cooling to form ferrite, which is the predominant phase in most steels at ambient temperatures. In order to produce fine grained steels with higher strength and toughness, grain refining additives which inhibit growth of the austenite at the normalising temperature provides a fine grained austenite with many more grain boundaries to act as nucleation sites for ferrite during cooling, thereby producing a finer ferrite grain structure.

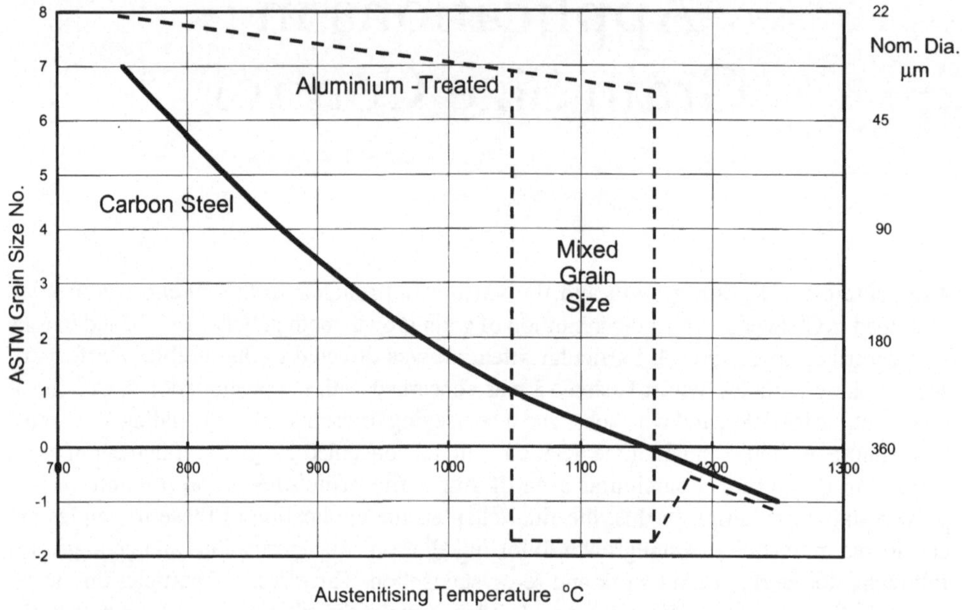

Fig 6.1 Effect of Temperature on the Austenite Grain Size.

Studies of austenite grain growth in plain carbon and aluminium-treated steels, (Bain and Paxton, 1961, Gladman and Pickering, 1967), show generally similar features. In plain carbon steels, i.e. steels without additions of second phase particles, held for about 1 hour at temperature, Figure 6.1, the grain size increases progressively from ASTM 7 (32µm diameter) at about 750°C, up to ASTM0 (360µm diameter) at 1250°C. In the case of aluminium-treated steels, the grain size remains fine at ASTM 7-8 (22-32µm diameter) at temperatures up to 1050°C or thereabouts, above which extremely large grains appear in the structure, giving rise to a mixed structure of the coarse grains (up to about 1 mm in diameter) with unchanged fine grains of about 30µm diameter. In transformable steels, observation of the high temperature grain structure is complicated by the allotropic change, which can produce a variety of micro-structural features at the prior austenite grain boundaries and in the prior austenite grain interiors. Such a structure is illustrated in Figure 6.2. Some indication of the fine grains can be seen at the extreme left of the figure, where the dark grain centres are revealed by a lighter (ferrite) rim. A single large dark grain can be seen occupying the bottom right quadrant of the figure and is at least 0.6mm in diameter. This mixed structure with either a single grain or very few large grains is characteristic of secondary re-crystallisation. At temperatures above about 1150°C, the grain size of the aluminium-treated steel reverts to that exhibited by the plain carbon steel, although in many cases the aluminium-treated, or 'fine-grained', steels shows slightly larger austenite grains than does the plain carbon steel. The temperature at which the very coarse grains appear is often referred to as the grain coarsening temperature, but it is important to note that this concept can only be used in relation to isochronal treatments. Increasing the length of the holding time reduces the temperature at which the coarse secondary re-crystallised grains appear. The grain size control theory can be applied to the

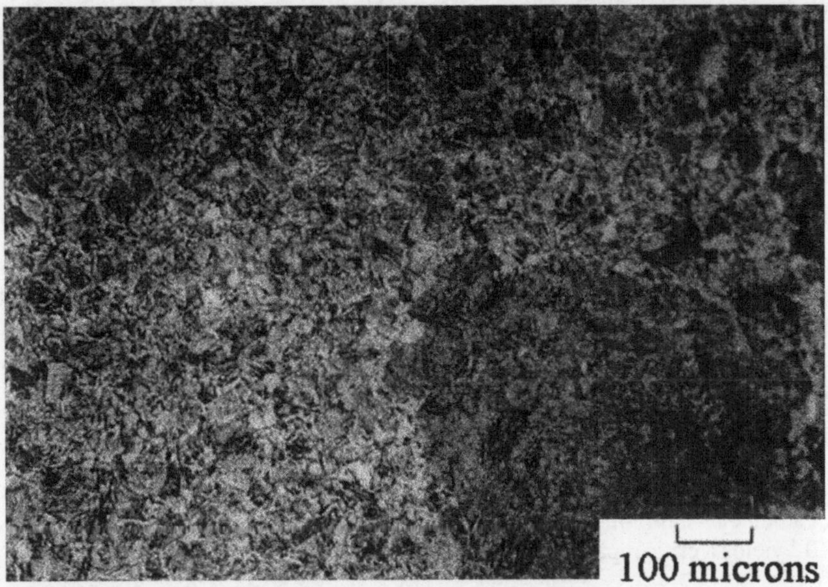

Fig. 6.2 A Secondary Re-crystallised Grain in a Low Carbon Steel; (see bottom right).

various features in Figure 6.4 to explain –

(a) The fine grain size of aluminium-treated steels below the grain coarsening temperature.

(b) The grain coarsening temperature and its slight time dependence.

(c) The cause of secondary re-crystallisation.

(c) The coarseness of the austenite grain size at temperatures above the secondary re-crystallisation range.

6.1.1 Fine Grained Normalised Steels

Given that the fine precipitated particles exist prior to the formation of the austenite grain structure on heating (which in fact they do), it is possible to define the particle size required to prevent any major grain coarsening, with an austenite grain size of, say, ASTM 8, for any given volume fraction of particles, or to specify the volume fraction required for any given particle size.

Consider Equation 4.11 (see Chapter 4) –

$$R_o = (1 - 4/3Z)r_c/f \qquad \text{(equ. 4.11)}$$

For a grain size of ASTM 8, the nominal diameter is 22μm and the radius of the grain would therfore be 11μm. Using a commonly observed value of grain size heterogeneity, Z = 5/3, then –

$$r_c/f = 5R_o = 55\mu m$$

If we then consider a 0.02wt.%Al-0.01wt.%N steel, the soluble aluminium content at 900°C (Figure 5.3) would be 0.006wt.% leaving 0.014wt.%Al as aluminium nitride. From stoichiometry this would give 0.021wt.% aluminium nitride, and hence, using the conversion factor of 0.0249 from Table 5.3, a volume fraction of 5.2×10^{-4}. The maximum particle size that would prevent coarsening under these conditions would therefore be –

$$r_c = 55 \times 5.2 \times 10^{-4} = 0.028\mu m$$

or 28nm

If the average particle radius is less than 28nm, then a fine grain size (ASTM 8) would be stable in that the (large) grains with the greatest size advantage would not be able to grow (unless of course the size advantage was somewhat greater than 5/3). If however, we consider the shrinkage of small tetrahedral grains, then Equation 4.16 shows that the critical size for the prevention of shrinkage is given by –

$$r_c = 1.935\ R_i f_v \qquad \text{(equ. 4.16)}$$

In section 4.2.2(b), it was pointed out that the maximum value of R_i is of the order of $0.27R_o$, so that substitution of the appropriate values of $R_i = 0.27 \times 11 = 2.97\mu m$, and $f_v = 5.2 \times 10^{-4}$ yields a critical particle radius of –

$$r_c = 1.935 \times 0.27 \times 11 \times 5.2 \times 10^{-4}\mu m$$
$$= 2.99nm$$

Thus any mean particle size above ~ 3nm would allow any tetrahedral grain in the structure to disappear. The disappearance of these small infilling grains would not however, cause any major instability of the granular structure and would only allow a slight adjustment of the fineness of the fine grain size. Clearly there are other geometrical possibilities, e.g. the shrinkage of five or six faced grains that would have intermediate critical particle radii.

Having established the critical particle sizes for growing tetrakaidecahedra, and for shrinking tetrahedra, the mean particle size that is present will control whether large grains will grow and small grains will shrink. Relatively crude experimental observations of the average size of aluminium nitride particles in a 0.02wt.% aluminium steel, heated to 900°C for 1 hour, (Gladman, 1963), indicated an equivalent radius of about 17nm. Thus, on the basis of the critical radii calculated, any small infilling tetrahedral should have disappeared, whilst the growth of the large tetrakaidecahedra (up to $Z = 5/3$), would still be inhibited. With such a difference between the observed (17nm) and critical (28nm) radii, only a substantial increase in the mean size of the particles, or decrease in the critical particle radius would allow grain growth to occur. Both of these events can be caused by a slight increase in temperature, which reduces the volume fraction of aluminium nitride, and therefore the critical radius, and allows more rapid growth of the particles by Ostwald ripening. Grain coarsening was observed in a sample of this steel heated to a temperature of 1000°C.

Larger additions of aluminium would give a larger volume fraction of aluminium nitride at 900°C (Figure 5.3), and would therefore inhibit a wider spectrum of grain growth events, e.g. would inhibit growth when the matrix grain size was smaller, and in the limit might even stop some of the largest tetrahedral grains from shrinking.

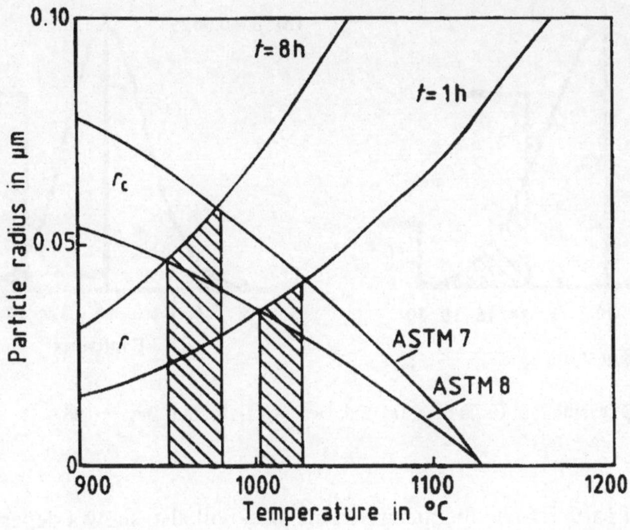

Fig. 6.3 Effects of Solubility and Particle Coarsening on the Grain Coarsening Temperature of an Aluminium Treated Steel.

6.1.2 THE GRAIN COARSENING TEMPERATURE

Figure 6.1 shows that a marked increase in grain size occurs at 1050°C, when some grains of up to about 1mm are observed. This can be described as the "grain coarsening temperature". However, the temperature at which the first major coarse grains are found is peculiar to the circumstances defined by the experiment. For example, the experimental observations depicted in Figure 6.1 relate to a 1hour treatment, and results obtained from an 8hour treatment would show a lower 'grain coarsening temperature', due to the progressive nature of Ostwald ripening. This is illustrated in Figure 6 (Gladman and Pickering, 1967), where the particle sizes after 1hour, and after 8 hours, at various temperatures are shown. Putting the critical particle size in the same diagram clearly reveals those areas where the actual particle size, r, is smaller than the critical particle size, r_c, and the onset of grain coarsening is inhibited, and also those areas where the actual particle size is larger than the critical particle size, and grain growth can occur. Clearly, the onset of grain growth occurs at lower temperatures as the holding time at temperature increases.

It should be noted that the particle size data used in Figure 6.3 is based on the very crude estimates of the effects of a 1 hour soak at different temperatures on the particle size, and in order to provide data for longer times, the Wagner equation (Equation 5.54) was used together with the assumption that r_0, the particle size at the start of the austenitising treatment, was small enough to be ignored. These aspects of Ostwald ripening will be considered in a later section.

Also shown in Figure 6.3, is the effect of matrix grain size. The finer the matrix grain size (R_0) the smaller will be the critical particle size, r_c, as shown by Equation 4.11. Consequently grain growth will commence at a smaller particle size, i.e. at a lower temperature or at a shorter time.

Thus the grain coarsening temperature, in addition to having a dependency on the

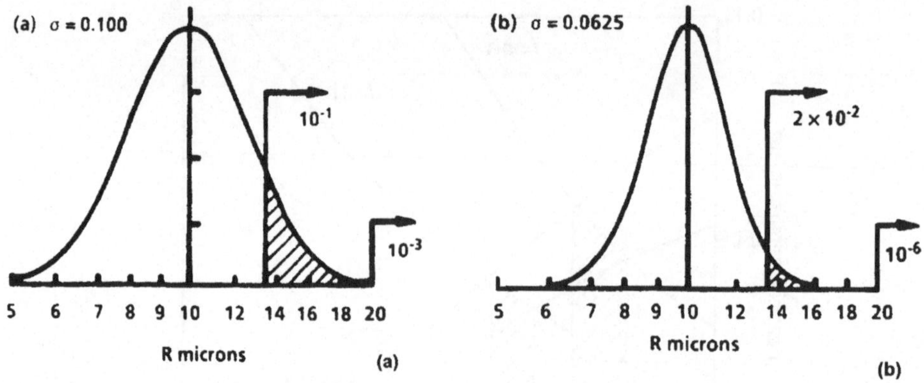

Fig. 6.4 Lognormal Distribution of Grain Size. (a). Std. Dev. = 0.1. (b). Std. Dev. = 0.06.

volume fraction of particles and the mean particle size, will also show a dependency on all of those factors which control the actual particle radius and the critical particle radius, i.e. the matrix grain size, the grain size heterogeneity, and the exposure time at elevated temperature. Also of importance is the prior thermal history of the material, as such treatments may result in a significant variation of the initial particle size prior to heating.

6.1.3 Secondary Recrystallisation

Normal grain growth involves the growth of all grains that have a positive driving force i.e. where growth will reduce the total energy of the granular system, Equation 3.19, and the shrinkage of all smaller grains that will similarly reduce the energy (e.g. see Equation 3.27). The shrinkage of the smaller grains and their ultimate disappearance will cause an increase in the mean grain size because there are fewer grains in a given volume of material, but the increase in size of each of the neighbouring grains will only be by a few percentage points, and is unlikely to be sufficient to cause one of the neighbours to grow. On the other hand, according to Equation 3.19, the energy change per unit area of boundary when any grain of radius R grows is given by –

$$dE_A/dR = (2/R - 3/2R_0)\gamma \qquad \text{(equ. 3.19)}$$

$$\text{and } dE_A/dR \text{ is negative only when } 3/2R_0 > 2/R$$

$$\text{or } R/R_0 = Z > 4/3$$

Thus, any grain with a radius greater than 4/3 times that of its neighbours will reduce the energy of the array when it grows. As indicated in earlier chapters, the distribution of grain sizes in an array of grains may be approximated by a lognormal distribution (Pande, 1992). Lognormal distributions of the grain radius are illustrated in Figure 6.4. Two distributions are shown, having different spreads i.e. standard deviations of $\sigma = 0.1$ in Figure 6.4(a), and $\sigma = 0.0625$ in Figure 6.4(b). The distributions are shown in terms of the grain radii (on a logarithmic scale) rather than in terms of the size advantage, Z. If it is assumed that any selected grain is surrounded by grains having an average radius which is the same as that of the overall mean grain size, then in the first case ($\sigma = 0.1$) one in ten grains would be qualified to grow in the normal grain growth process, i.e. all the grains having a radius greater than 133μm. If the standard deviation is reduced to 0.0625, then only one in 50 grains would be able to grow. Such arguments are only semi-quantitative as a number of grains having the same diameter would show a distribution of Z values depending upon their location with respect to having smaller and larger grains as neighbours. Also, if the distribution is of a lognormal type, the mean logarithm of the radii would correspond with the geometric mean grain radius rather than the arithmetic mean radius. However, in the distributions shown, the differences between the two types of mean are negligibly small. A normal distribution of grain sizes would also yield similar results. It can be concluded, therefore, that a significant proportion of grains would be eligible to grow (i.e. would have the necessary size advantage) giving rise to a normal grain growth process. It is important to realise that when the average grain radius is doubled, the grain volume is increased by a factor of eight, and about 80% of the grains will have disappeared.

The essential feature of normal grain growth is that many grains are growing, and these grains will ultimately become neighbours, and the whole question of size advantage and competitive growth will start again, but in a matrix of much larger grains. The larger grain size of the matrix, R_0, in a steady state distribution of sizes (where only the scale of the distribution changes), will reduce the rate of growth by reducing the rate of change of energy. If we rewrite Equation 3.19 as –

$$dE_A/dR = (2/Z - 3/2)\gamma/R_0$$

then, given that Z is sensibly constant, the rate of growth of a grain is inversely proportional to the matrix grain size.

Now consider the situation where an array of second phase particles has been introduced to inhibit grain growth completely, as in the low temperature region of Figure 6.1 Due to the progressive ripening of the particles, it follows that at some stage, the average particle size, at the given volume fraction, will become ineffective in pinning the grain with the largest size advantage, Z. All other grains with smaller values of Z will still be effectively pinned. Thus we have a single grain growing into an unchanged matrix, i.e. Z increasing whilst R_0 is unchanged. From the equations for the driving forces and pinning forces for the growing tetrakaidecahedron, Equations 3.19 and 4.5 respectively, we can write –

$$dE_A/dR = (2/Z - 3/2)\gamma/R_0 + 3\gamma f_V/2r \qquad \text{(equ. 6.1)}$$

– for the single growing grain, the rate of decrease in energy increases, i.e. dE_A/dR becomes more negative, as Z increases when R_0 is constant, R_0 being defined as the matrix grain size into which the single grain is growing. Clearly, after an interval of time in which the single grain has grown with an increasing rate of growth, the particles will have coarsened further to allow a second grain to start to grow, i.e. The grain with the second largest value of Z .

The relationship between the average particle radius and the size advantage, Z, necessary for the grains to grow can clearly be seen in Figure 4.6 for any given levels of volume fraction of particles and matrix grain size. For example, when the mean particle radius reaches a value of $3R_0f_V$, grains with a size advantage of Z = 2 will be released. The particles have to grow by a further 33% (to $r = 4R_0f_V$) before grains with a size advantage of Z = 1.75 are released. Clearly, the grain size distribution will control the numbers of abnormal grains. Some indication of the frequencies could be obtained from a three-dimensional distribution of grain radii shown in Figure 6.4. If it is assumed that the average

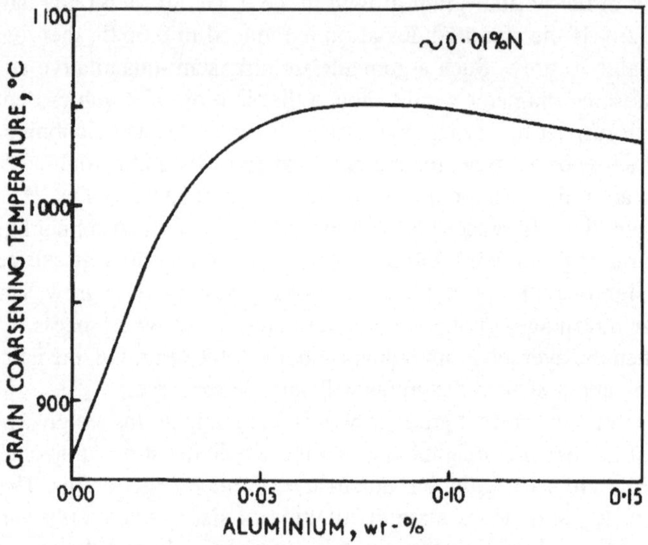

Fig. 6.5 The Effect of Aluminium Content on the Grain Coarsening Temperature.

grain size is approximately 10μm, then the incidence of grains with a radius of 20μm or more, i.e. Z = 2 or more, is 1 per thousand grains in Figure 6.4(a), and 1 per million grains in Figure 6.4(b). Figure 6.4 is schematic but shows that normal grain growth and abnormal grain growth can occur with these distributions given the appropriate absence or presence of adequate pinning forces. In fact the values of Z derived from grain coarsening data, e.g. Z values of 1.5 – 1.8, suggests that the initial grain size distributions are quite narrow.

The abnormal growth process permits the growth of several grains, but the selective nature of grain release means that there are few growing grains, and this growth will only stop when the few grains collide to give a huge (millimetric scale) grain size, or when a single grain consumes the array. At any stage prior to this, the microstructure consists of huge grains growing into an unchanged fine grained matrix. As described earlier, this is the process known as secondary re-crystallisation or abnormal grain growth, the result of which is shown in Figure. 6.2.

The mechanism of abnormal grain growth in the case of a pinned structure, where the particles are growing, differs significantly from that in thin foils where the pinning force per unit area of boundary is provided by thermal grooving. This is overcome by the driving force when the matrix grain size is small, but the driving force is reduced as normal grain growth occurs, until only the grains with a large size advantage can grow, as shown in Figure 4.13. In this case secondary re-crystallisation (abnormal grain growth) follows the normal grain growth process (see Sections 4.3.1 and 4.3.2).

6.1.4 The Grain Size at Very High Temperatures in the Austenite Range

At temperatures above the range where abnormal grain growth and mixed grain sizes are seen, Figure 6.1, the grain size is shown to be of the same order as that seen in a steel without deliberate grain refining additives. This will of course only be seen if relatively

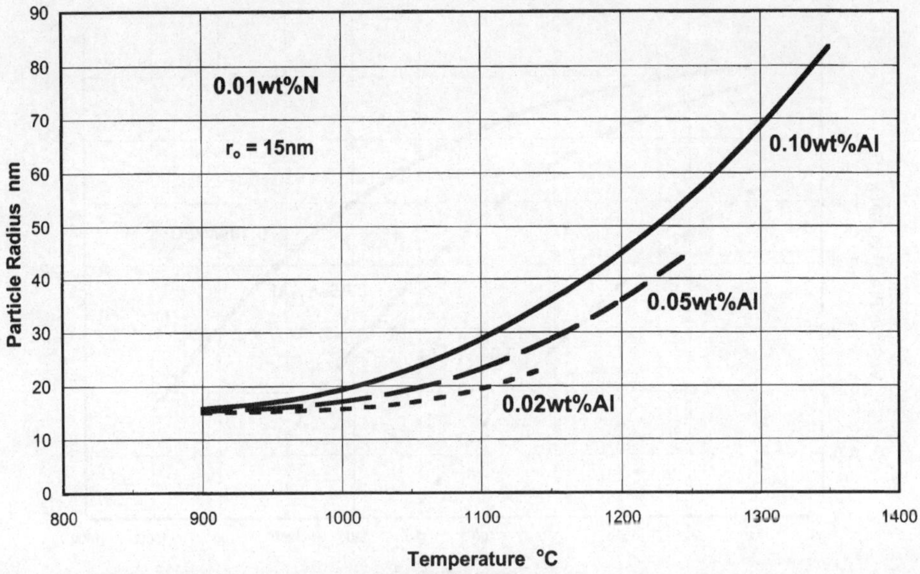

Fig 6.6 The Effect of Solute Content on the Ripening of AlN.

small samples are rapidly heated to the various temperatures in the austenite range and then held isothermally to allow grain growth to occur. Under these circumstances, the volume fraction of aluminium nitride is dramatically reduced, as shown in Figure 6.3, and the rate of ripening increases. Thus, normal grain growth can occur from the size formed immediately after transformation, (ASTM 8 or thereabouts) to that seen in plain carbon steels (ASTM 0 to ASTM –1). These very coarse grain sizes may well reflect the pinning effect of inclusions such as oxides and sulphides commonly found in steels.

However, if we consider the grain growth processes during slow continuous heating to the higher temperatures, it is clear that abnormal grain growth will occur on passing through the critical temperature range, and the abnormal grains will persist at the higher temperatures. This phenomenon has frequently been observed in large forgings and during the heat treatment of large castings.

6.2 THE EFFECTS OF ALUMINIUM CONTENT ON GRAIN COARSENING

The effect of increasing the aluminium content on the grain coarsening temperature of austenite has been known for many years (Chatterjee and Nijhawan, 1957). Increasing the aluminium content from about 0.01wt.%Al up to 0.05wt.%Al causes a progressive increase in the grain coarsening temperature of steels containing about 0.01wt.%N. Further increasing the aluminium content has little further effect, and, at levels above 0.10wt.%Al, may actually reduce the grain coarsening temperature, Figure 6.5. There are two factors, which may contribute to this effect. Firstly, the rate of coarsening can be affected by the solute content, and secondly, the greater amounts of aluminium nitride formed at higher temperatures may allow more pronounced coarsening of the particles during prior thermal treatments.

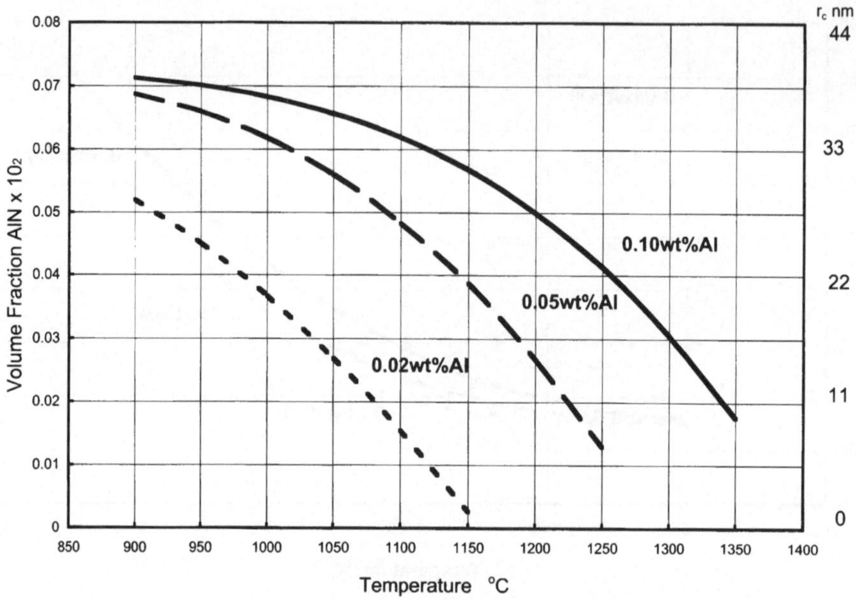

Fig 6.7 Effect of Aluminium Content on the Dissolution of AlN and the Critical Particle Size, r_c.

6.2.1 The Solute Content

According to the Wagner equation for bulk diffusion –

$$r^3 - r_0^3 = 8D\sigma\Omega St/9RT \qquad \text{(equ. 5.54)}$$

– the rate of change of the mean particle volume is directly proportional to the atom fraction of the rate limiting solute. This effect is illustrated in Figure 6.6 for steels containing 0.01wt.%N with 0.02, 0.05, and 0.10wt.%Al. The data given in Chapter 5 for solubility, diffusion coefficients, and molar volume, were used, and an interfacial energy of 0.8Jm^{-2}, was assumed. The mean particle radii attained after one hour at various temperatures are shown. It should be noted that the Wagner equation requires that there is an equilibrium amount of aluminium nitride already present, and that this is in the form of particles of radius, r_0. For the purpose of the present illustration, an initial particle radius of 15nm was assumed, this being in accord with experimentally observed particle sizes in samples exposed for 1hour at 900°C. The main features of Figure 6.6 are –

(i) There is very little change in the particle radius in 1hour at 900°C irrespective of the aluminium content of the steel.

(ii) Increasing the temperature markedly increases the particle radius, reflecting the increase in solute content with temperature, and also the increase in diffusion rate of the solute.

(iii) At a given temperature, increasing the aluminium content of the steel increases the soluble aluminium level, Figure 5.3, and causes an increase in particle radius for a given exposure time.

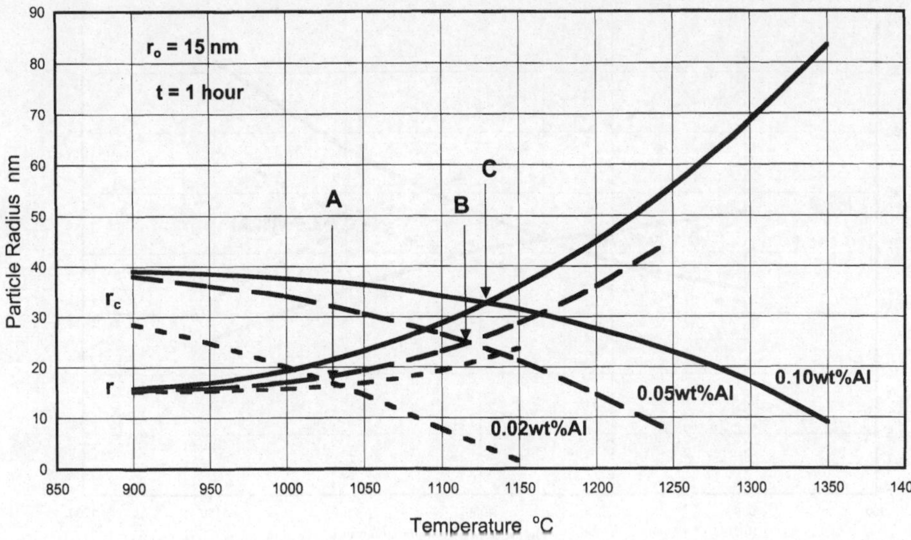

Fig 6.8 Effect of Aluminium Content on Grain Coarsening Temperature.

The significance of this increase in particle radius with increasing aluminium content can only be judged in conjunction with the corresponding change in the volume fraction of aluminium nitride, and the consequent effect on the critical particle size as discussed in Section 6.1.1. The effect of aluminium content on the volume fraction of aluminium nitride at various temperatures is shown in Figure 6.7. Increasing the aluminium content of the steel increases the volume fraction of aluminium nitride in accord with Figure 5.3. Following the example shown in Section 6.1.1, the critical particle size r_c is given by –

$$r_c = 55f_v \ \mu m$$
$$\text{or} = 55{,}000f_v \ nm$$
$$\text{– when } Z = 5/3, \text{ and } R_0 = 11\mu m \text{ (ASTM8).}$$

This allows a direct scaling between the volume fraction and the critical particle size, as shown on the right hand ordinate of Figure 6.7. The significance of the enhanced particle coarsening due to steel composition and the increase in dissolved aluminium level may be seen from the locus of the crossover points between the calculated particle sizes and the critical particle sizes as shown in Figure 6.8. The expected grain coarsening temperature (when $r > r_c$) increases markedly on increasing the aluminium content from 0.02wt.% (Point A) to 0.05wt.% (Point B), but shows only a very slight increase on further increasing the aluminium content to 0.10wt.% (Point C).

This reduction in the effect of aluminium due to the solute affected Ostwald ripening can be appreciated by considering the crossover points between the critical particle size curves and a single Ostwald ripening curve, say for the 0.05wt.%Al steel in Figure 6.8. The solute-affected ripening reduces the crossover points and therefore the predicted grain coarsening temperature. Although the predicted grain coarsening temperatures are not exactly those shown in Figure 6.5, the absence of precise data on the value of r_0 and the

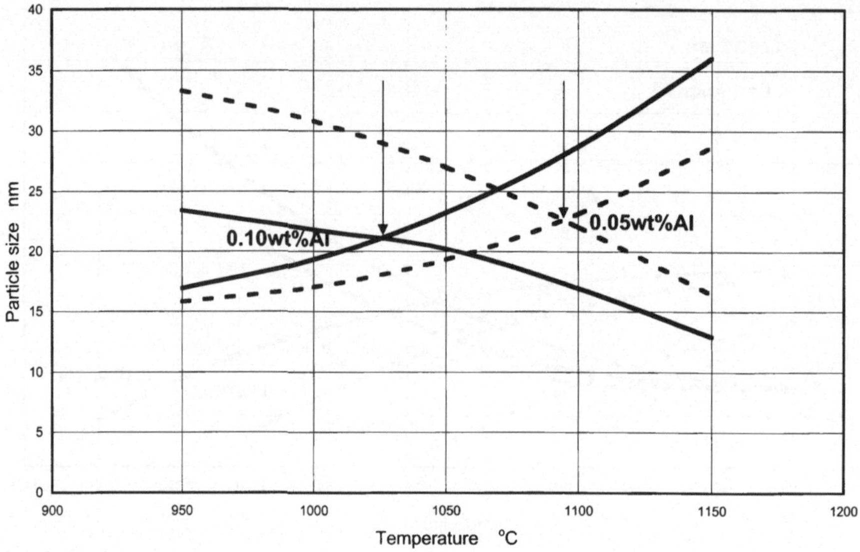

Fig 6.9 Effect of Particle Pre-coarsening on the Grain Coarsening Temperature.

value of the diffusion coefficient for aluminium in austenite, together with the complete omission of possible effects from prior processing, make the closeness of the estimated grain coarsening temperatures quite remarkable.

6.2.2 Effects of Prior Treatments

The relatively high thermal stability of aluminium nitride in high aluminium steels (above about 0.05wt.%) makes the steels sensitive to prior treatments. For example, suppose that the steels are soaked at 1250°C for 2 hours prior to hot rolling. In the 0.02wt.%Al steel, all of the aluminium nitride is dissolved, as shown in Figure 6.7, and would be available to form precipitates as discussed in the previous section. In the case of the 0.05wt.%Al, the volume fraction of aluminium nitride which remains out of solution at 1250°C is 1.3×10^{-4}, and this will undergo Ostwald ripening. Using Equation 5.54 and the same data that was used in the previous section suggests a particle size of 57nm. The size of this undissolved fraction of coarse particles is already in excess of the critical particle size, so that the subsequent grain coarsening characteristics of this steel will very much depend on the fraction dissolved at 1250°C, and the form in which it precipitates. Under extremely slow cooling conditions, it is possible that further precipitation would take place on the existing particles and would hardly change the pinning force of the array. However, within the times and temperatures of interest in hot working, substantial precipitation from austenite on widely dispersed particles is unlikely, (Honer and Baliktay, 1977; Konig, Schultz and Ulmer, 1961). Aluminium nitride shows a C-curve for precipitation on direct cooling from high temperatures, with a maximum rate of precipitation at 1100°C, although even here the precipitation took some 6 hours. Relatively rapid precipitation can occur during hot deformation when nucleation in the supersaturated solid solution occurs on dislocation arrays or sub-grain boundaries, Figure 5.15. Rapid precipitation can also take place during

Table 6.1 Calculation of the Critical Particle Radius, r_c, with Mixed Particle Sizes.

Temp. °C	0.05wt%Al			0.10wt%Al		
	f_v	f_v-f_u	r_c nm	f_v	f_v-f_u	r_c nm
950	0.66x10^{-3}	0.53x10^{-3}	33.3	0.70x10^{-3}	0.29x10^{-3}	23.4
1000	0.62x10^{-3}	0.49x10^{-3}	30.8	0.68x10^{-3}	0.27x10^{-3}	21.8
1050	0.56x10^{-3}	0.43x10^{-3}	27.0	0.66x10^{-3}	0.25x10^{-3}	20.2
1100	0.48x10^{-3}	0.35x10^{-3}	22.0	0.62x10^{-3}	0.21x10^{-3}	16.9
1150	0.39x10^{-3}	0.26x10^{-3}	16.4	0.57x10^{-3}	0.16x10^{-3}	12.9

transformation or in the ferrite phase, Figure, 5.16. In both cases the accelerated precipitation is associated with the formation of an array of much smaller particles. Thus, a bi-modal particle size distribution exists, arising from the undissolved fraction, f_u, of aluminium nitride particles of a size, r_u, and the newly formed fraction, f-f_u, of aluminium nitride particles which are subject to Ostwald ripening. Although there has been little attention given to the ripening process in bimodal distributions of particles, closely spaced fine particles between widely spaced coarse particles could exhibit the same coarsening characteristics as the single array considered in Section 6.2.1. This could be considered as an optimum case when undissolved coarse particles are present, but inactive. Under these conditions, the grain coarsening behaviour with mixed particle sizes (Section 4.2.5) can be invoked. Consider Equation 4.29 –

$$R_0 = (1 - 4/3Z)/\Sigma(f_i/r_i)$$

where f_i is the volume fraction of the 'i'th class
and r_i is the particle radius of that class.

We now have two classes of particle, so that –

$$(f - f_u)/r_c + f_u/r_u = (1 - 4/3Z)/R_0 \qquad \text{(equ. 6.2)}$$

the suffix u relating to the undissolved aluminium nitride at the soaking temperature prior to hot working. f is the total aluminium nitride fraction present after re-austenising at various (lower) temperatures. With a grain radius, R_0 of 11μm, and a grain size heterogeneity factor of 5/3, rearranging Equation 6.2 gives –

$$r_c\,(\text{nm}) = (f - f_u)/(1/55000 - f_u/r_u) \qquad \text{(equ. 6.3)}$$

The critical radius, r_c, now relates to the average size of the newly formed precipitate. The results are shown in Table 6.1 for the 0.05wt.% and 0.10wt.%Al steels. The critical particle radii are now smaller than those shown in Figure 6.7, and will therefore result in lower grain coarsening temperatures, as indicated in Figure 6.9. The critical radii of the 0.10wt.%Al steel are now less than those of the 0.05wt.%Al steel, due to the greater volume fraction of the even larger undissolved aluminium nitride particles, (also included in Table 6.1). The effects of the pre-coarsened particles gives a clear indication that this effect can explain a decrease in grain coarsening temperature with increasing aluminium content at the higher levels of aluminium. Although there is no direct correspondence between the experimental curve shown in Figure 6.5 and the calculated values of Figure 6.9 (with respect to the grain coarsening temperatures and the optimum aluminium content), there are obviously several important parameters in grain coarsening that are frequently

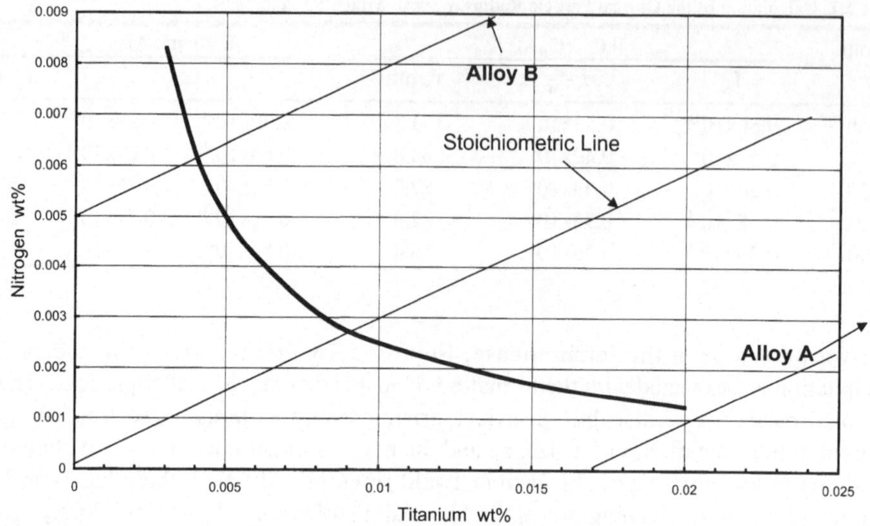

Fig 6.10 Solubility of Titanium Nitride at 1350°C.

unavailable from published experimental work. These variables have real physical meaning, and are real physical quantities, but may be difficult to measure, or may simply have remained unrecorded. These include –

(i) The grain size advantage (or heterogeneity factor) Z. The driving force for grain growth is highly sensitive to small changes in Z. For example, with Z = 5/3 as used above, the critical particle radius, given by Equation 4.11, is $5R_0f$. Reducing Z to 1.5 increases the critical particle radius to $9R_0f$, whilst an increase in Z to 2.0 reduces the critical particle radius to $3R_0f$. These values could dramatically change the predicted grain coarsening temperature, as a quick check with Figure 6.8 will show. However inconvenient this sensitivity is, size advantage is a physical reality, and any theory would be seriously suspect without it.

(ii) r_o values in the Wagner Equation can only be ignored if the particle sizes are very many times larger than r_0. Unfortunately, in the grain refinement of steel, where every effort is made to minimise the amount of costly additives, this is not necessarily the case, and the selection of r_0 values in the present work was largely based on experimental estimates of particle size, r, after a 900°C treatment, where Ostwald ripening of aluminium nitride particles is very slow, and, also the solute effect is minimal (Figure 6.6).

(iii) The prior thermal history of the material, particularly in relation to the pre-coarsening of some or all of the grain refining particles, can be very important. Unfortunately details of soaking times and temperatures, especially in steels with stable particles, are rarely disclosed, making estimates of the volume fraction and size of pre-coarsened particles impossible.

With these shortcomings in mind, making a better fit between the experimental and theoretical grain coarsening temperatures would be relatively easy, but would not give any further understanding.

6.2.3 General Observations on Aluminium Treatment

The effects of aluminium content on the rate of particle coarsening, and on the presence of coarse undissolved particles, are contributory factors in the lack of effectiveness when aluminium contents are increased at levels above about 0.05wt.% aluminium, with respect to both the grain coarsening temperature and the size of the austenite grain that can be stabilised during normalising.

The example of soaking for hot rolling given in Section 6.2.2, reflects the days when steels were manufactured as ingots, stripped from the mould, stacked, and subsequently charged to a soaking furnace for several hours. Nowadays, when continuous casting has almost completely replaced ingot production, and the continuously cast product may be direct rolled with little more than a token temperature equalisation treatment, the processing variables may have less significance, in that treatment times are much reduced.

Other changes in steel manufacture also have implications for the detailed application of grain size control. In Sections 6.1 and 6.2, the examples have concerned the use of aluminium nitride, and a nitrogen level of 0.01wt.% has been assumed, bearing in mind that the different steel-making process can give different nitrogen levels. For example, in the 'old' manufacturing processes, nitrogen levels were typically –

Open Hearth Process 0.007wt.%N
Electric Arc Process 0.010wt.%N
Bessemer Process 0.015wt.%N

– and grain refining additions and practices involving nitrides would be centred on the appropriate nitrogen level, even though nitrogen levels could be adjusted by using high nitrogen additives. Bulk steel-making is gradually moving to the AOD process, where much lower nitrogen levels are not uncommon (about 0.005wt.% or less), so that nitrogen control will be important, or alternatively, grain control may be practiced with alloy carbide particles.

Whilst the examples selected so far have been based on aluminium steels, the principles involved are applicable to any of the grain refining particles, and both vanadium nitride particles and niobium carbo-nitride particles have been used to control the prior austenite grain size of normalised steels. Use of the appropriate data given in Chapter 5 for the solubilities, diffusivities, etc. allows steel compositions to be specified to give resistance to grain coarsening. It is perhaps advisable to reiterate the warning that the use of Equations 4.11, 4.23, 4.29, and 4.30 merely describe the borderline between grain coarsening and the inhibition of grain growth for the different types of particle distribution, e.g. random, grain boundary, mixed sizes. If the actual mean particle size is less than the critical particle size, then grain growth will be inhibited, and the array of particles would be capable of inhibiting grain growth for an even finer grain size. The fact that the particles are capable of pinning a finer grain size, does not

Table 6.2 Comparison of the Solubility Products of Aluminium and Titanium Nitrides at 1350°C.

Nitride	k_s (1350°C)	Source
AlN	7.2×10^{-4}	Leslie et al, 1954; Irvine et al, 1967
TiN	2.5×10^{-5}	Matsuda and Okamura, 1978
TiN	3.7×10^{-6}	Narita, 1975

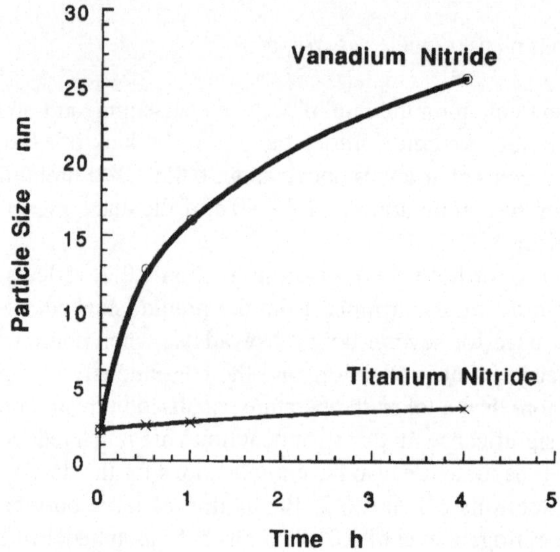

Fig. 6.11 Calculated Coarsening Rates of TiN and VN at 1100°C.

make the grain size any finer, as the establishment of the austenite grain size in normalising treatments arises from the nucleation and growth of austenite during the allotropic transformation on heating. If this transformed austenite grain size has a radius of 11μm (ASTM 8), then the presence of particulate arrays capable of pinning grains of radius 8μm (ASTM 9) or 4μm (ASTM 11) would not change the grain size. Both arrays would prevent the growth of grains of radius 11μm. On the other hand, if the particulate array could only stabilise an array of grains of 16μm radius (ASTM 7), then grain growth would commence.

6.3 TITANIUM NITRIDE TECHNOLOGY

Steels with relatively low titanium contents (with respect to nitrogen) had been shown to contain very fine particles of titanium nitride (George and Irani, 1968), and further, these fine particles proved to be very resistant to Ostwald ripening. The reasons for this are now clear. The Wagner equation for bulk diffusion controlled Ostwald ripening (Equation 5.54), includes a term for the atom fraction of the rate limiting solute.

$$r^3 - r_0^3 = 8D\sigma\Omega St/9RT$$

In Section 6.2.1, attention was drawn to the effects of the solute content in the case to aluminium treated steels. In the case of titanium nitride, however, the solubility in austenite is far lower than that of aluminium nitride. Table 6.2 compares the solubility products for the two nitrides in austenite. The solubility data for titanium nitride from two sources are shown (Matsunda and Okamura, 1978; Narita, 1975). The solubility product for TiN at 1350°C is between 28.0 and 200 times less than that of aluminium nitride, depending on the data source chosen, so that the dissolved titanium will be between 28.0 and 200 times less than the dissolved aluminium, for any given soluble nitrogen level. Given that the relative atomic masses of titanium and aluminium are 48 and 27 respectively, the level of

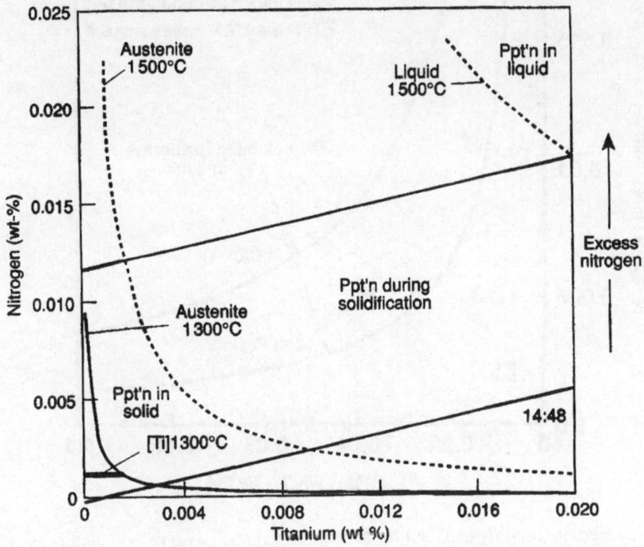

Fig. 6.12 Solubility of TiN in Austenite and Liquid Steel.

solute, S, (where S is expressed as an atom fraction) for titanium is between 50 and 350 times less than that for aluminium, so that the rates of change of mean particle volume should differ by this same factor. As the solubility of titanium nitride is considerably below those of any of the micro-alloy carbides and nitrides, the exceptionally high resistance to particle coarsening is not surprising.

The absolute level of particle coarsening and its dependence upon the rate limiting solute can also be controlled by alloy chemistry, as indicated in Figure 6.10. By definition, the solubility product is given by –

$$k_s = [Ti][N] \qquad \text{(equ. 6.4)}$$

where $[T_i]$ is the dissolved titanium content (wt.%)
and [N] is the dissolved nitrogen content (wt.%).

The dissolved levels of these two elements are thus interdependent, and at any given temperature are dependent upon the composition of the alloy with respect to the stoichiometric ratio. Even when using the higher solubility levels of Matsuda and Okamura (1978), for stoichiometric alloys, the [Ti] and [N] levels are 0.0093 and 0.0027 wt.% respectively. These values compare with 0.037 and 0.019 wt.% for aluminium and nitrogen respectively, and indicate the effect of the solubility product on Ostwald ripening after corrections have been made to give atom fractions (1.085×10^{-4} and 7.674×10^{-4}) for titanium and aluminium respectively.

As shown in Figure 6.10, the titanium in solution can be varied considerably by changing the titanium and nitrogen contents to give a hyperstoichiometric titanium ratio (alloy A) or a hypostoichiometric titanium ratio (alloy B). Alloy A shows an excess of 0.0171wt.%Ti over the stoichiometric ratio, which corresponds to a nitrogen deficiency of 0.005wt.%N. In this alloy, the soluble titanium level at 1350°C is 0.0212wt.%. Alloy B shows an excess of 0.005wt.%N over the stoichiometric ratio, which corresponds with a

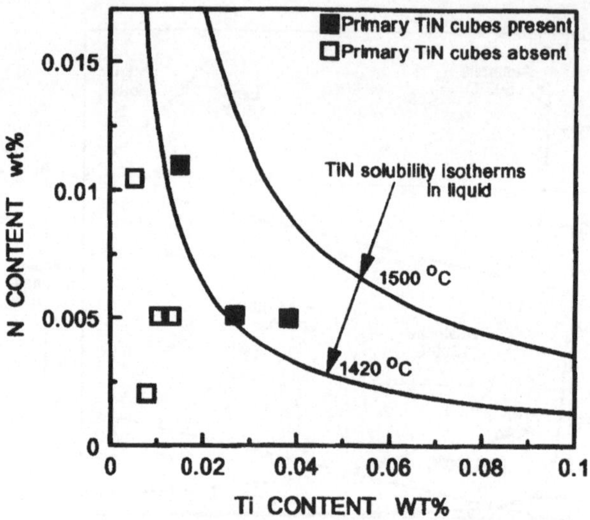

Fig. 6.13 The Occurrence of Cuboids in Titanium Steels.

titanium deficiency of 0.0171wt.%Ti. In this alloy the soluble titanium level at 1350°C is 0.0034wt.%Ti. Alloy compositions with hypostoichiometric titanium compositions can thus show very low solute levels. As shown in Figure 6.10, low titanium levels in equilibrium with titanium nitride can be attained by reducing the titanium contents,or by increasing the nitrogen content. The importance of substitutional diffusion relative to interstitial diffusion, being some 2 x 10^{-4} times slower, has already been discussed in Chapter 5, (see Table 5.7). It should be noted that according to the solubility data given by Narita (1975) the reduced solute effects are expected to be even more pronounced than those discussed above and shown in Figure 6.10 (cf Figure 5.8). Some indication of the reduced solute effects can be obtained from a comparison of the calculated coarsening rate of titanium nitride with that of vanadium nitride, Figure 6.11, which shows a level of particle coarsening more typical of the micro-alloy nitrides. The very slow coarsening rate calculated for titanium nitride is very much in accord with the lack of particle growth in hypostoichiometric alloys observed by George and Irani (1968).

Having extolled the virtues of low solute contents in order to restrict the extent of Ostwald ripening severely, the very low solubility of titanium nitride brings its own

Table 6.3 Ostwald Ripening of TiN in a 0.02wt%Ti-0.01wt%N Steel.

Temp. °C	k_s Matsuda and Okamura, 1978.	Atom Fraction Titanium, S	Mean Particle Radius, r*
900	3.16x10^{-7}	0.88x10^{-6}	10.001
1000	1.09x10^{-6}	3.00x10^{-6}	10.02
1100	3.11x10^{-6}	8.29x10^{-6}	10.29
1200	7.74x10^{-6}	19.40x10^{-6}	12.29
1300	1.71x10^{-5}	38.80x10^{-6}	19.10

**Initial mean particle radius,* r_o *= 10nm was assumed.*

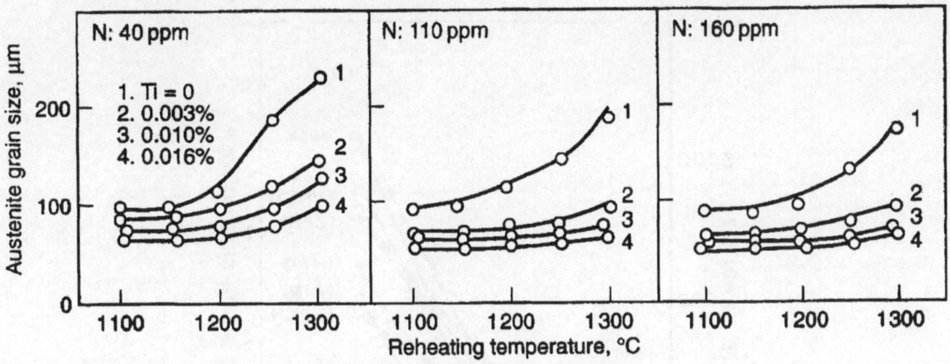

Fig. 6.14 Effects of Titanium and Temperature on the Austenite Grain Size. (After Osuzu).

attendant problems. Additions of titanium at typical micro-alloying levels cause some of the titanium nitride to precipitate in the liquid prior to the onset of solidification, Figure 6.12. This is followed by extensive precipitation during solidification, and finally precipitation in the solid state. The importance of these precipitation processes is that the particles formed in the liquid are very large and take the form of cuboids with an edge length of up to 5µm, whilst those formed during solidification are usually smaller by still large by grain refining standards, having edge lengths of 0.5 – 1.0µm. The particles precipitated in the solid state are much finer, having equivalent radii of the order of 10 – 15nm. It should be pointed out that Ostwald ripening can also occur in the liquid state and that the sizes given above represent upper bound levels for ingot cast materials. With modern steel-making methods, rapid metal transfer, and continuous casting, the particles formed in the liquid and during solidification should be much smaller than the sizes quoted.

From the above, it is clear that the very significant variations in size must be taken into account when using titanium nitride as a grain refining agent. If we select a steel having 0.02wt.%Ti and 0.01wt.%N, then only 0.0093wt.%Ti will be in solution at 1500°C, i.e. at the end of solidification. During solidification, some partitioning of the titanium and nitrogen may occur (Gladman, 1992; Adrian, 1995), which may change the balance of titanium and nitrogen in the solid, but with very low titanium and nitrogen levels, this effect may be ignored, so that the dissolved titanium and nitrogen levels in the austenite can be obtained by the usual method from Figure 6.12. The nitrogen in solution at this stage is 0.0069wt.%, and precipitation in the solid state will ultimately

Table 6.4 Borderline Austenite Grain Radius in a 0.02wt%Ti-0.01wt%N Steel.

Temp. °C	Fine Particles		Coarse Particles		$\Sigma(f_i/r_i)$ µm-1	R_o µm*
	f_1	r_1 nm	f_2	r_2 nm		
900	$1.79x10^{-4}$	10.001	2.07x10-4	1000	0.0181	7.9
1000	$1.75x10^{-4}$	10.02	—	—	0.0177	11.3
1100	$1.66x10^{-4}$	10.29	—	—	0.0163	12.3
1200	$1.48x10^{-4}$	12.29	—	—	0.0122	16.6
1300	$1.16x10^{-4}$	19.10	—	—	0.0063	31.0

**A value of Z = 5/3 was assumed. Grains with a larger radius R_o than that shown will not grow.*

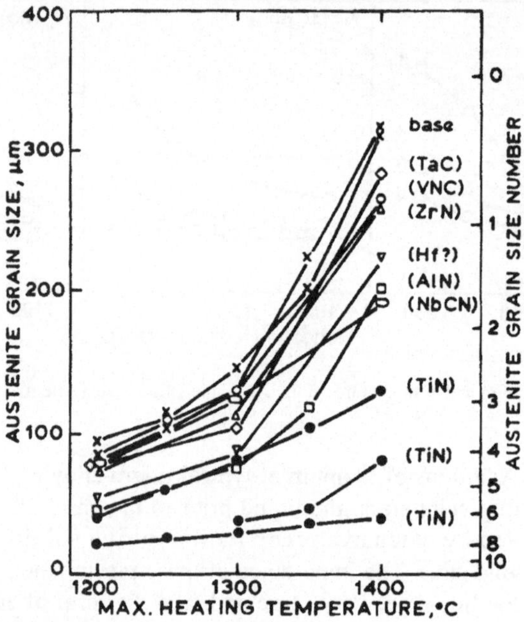

Fig. 6.15 Effect of Grain Refiners in the Heat Affected Zones of Weldments.

leave an excess pf nitrogen in solution (as would be expected from the hypostoichiometric titanium content of the alloy) of about 0.0042wt.%N. The outcome of these processes is that about 0.02 – 0.0093 = 0.0107wt.%Ti will be in the form of TiN particles precipitated during solidification, having a size of about 1µm, whilst the remainder of the titanium (0.0093wt.%Ti) will be in the form of fine particles, Figure 6.13 (Priestner, Zhou, and Ibraheem, 1997). The volume fraction of fine particles will decrease with increased reheating temperature, whilst Ostwald ripening of the particles will increase. Table 6.3 shows the effect of reheating temperature on the solubility product, the rate limiting solute, S, (the atom fraction of titanium) and the mean particle radius of the fine particles, calculated from the Wagner equation (Equation 5.54). An initial particle radius, r_o, of 10nm was assumed. The mean particle radius after 1hour exposure shows very little coarsening at temperatures below 1300°C, and even at 1300°C, the particle radius is less than twice that found after lower temperature exposures at 900°C. The effects of maintaining such a fine particle size, and the volume fraction of titanium nitride, on the matrix grain size was calculated using the equation for mixed particle sizes (Equation 4.29) –

$$R_o = (1 - 4/3Z) / \Sigma(f_i/r_i)$$

If a value of Z = 5/3 is assumed, then the equation becomes –

$$R_o = 1/5\Sigma(f_i/r_i) \qquad \text{(equ. 6.5)}$$

There are two classes of particle, i.e. those that were precipitated during solidification with a particle radius of about 1µm, and those precipitated in the solid, which are much finer. Table 6.4 shows the volume fractions and sizes of the fine particles and the calculated

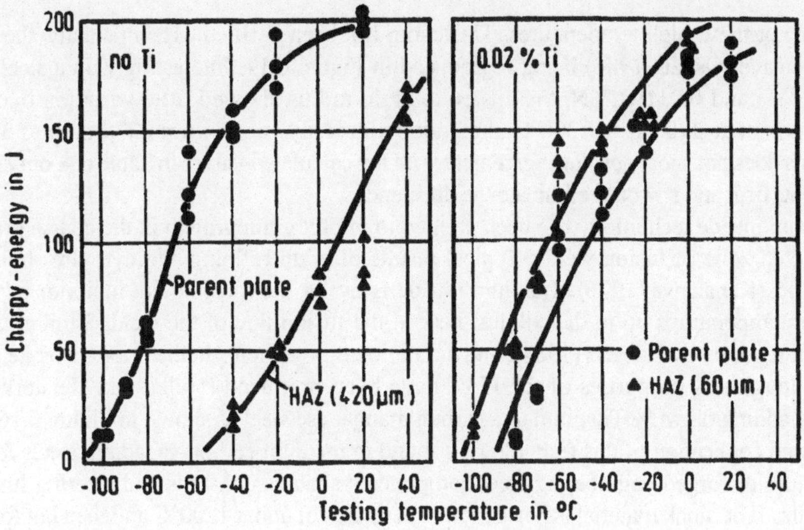

Fig. 6.16 Effect of Titanium on the Grain Size and Toughness of Steel.

volume fraction of large particles. It can be seen that the matrix grain sizes calculated from Equation 6.5 are finer than ASTM7 at all temperatures up to 1300°C, so that a reaustenitisation treatment that results in the formation of an initial austenite grain size of ASTM7 or thereabouts would not show any grain growth.

It should be emphasised that the above treatment is extremely simplistic. There are several factors which have been omitted or assumed.

(i) The question of elemental partitioning between solid and liquid has been ignored. This problem has been discussed by Pickering (1997). The effects of other alloys on the liquidus temperature and on the solidification temperature range were also considered. (See Figure 6.13).

(ii) The solidification process in low carbon steels commences with the formation of the b.c.c. δ-ferrite. Austenite is then formed by a peritectic reaction between the liquid and the δ-ferrite. It is only in the higher carbon steels that austenite forms directly from the liquid (0.4 – 0.5wt.%C). This has been ignored.

(iii) Various important parameters have been assigned "reasonable" values, e.g. the grain size heterogeneity factor (or size advantage), Z, was assumed to have a value of 5/3, and the fine particle radius prior to heating, r_o, was assumed to be 10nm.

With different values of Z, r_o, and by introducing partitioning, the austenite grain radii shown in Table 6.4 could be changed significantly. However, what is being claimed in this section on 'titanium nitride technology' is that the very low solubility of titanium nitride in austenite gives rise to very low titanium contents in the matrix, and that the use of hypostoichiometric titanium levels further accentuates this effect, so that Ostwald ripening, which depends on the flux of titanium atoms in the matrix, is extremely slow. The result is that the extremely stable particles maintain a very small size up to very high temperatures, and therefore inhibit austenite grain coarsening at these high temperatures. This claim has been verified experimentally, the most comprehensive (Osuzu, 1986) being reproduced in Figure 6.14. Interestingly, in a steel with 0.016wt.%Ti and 0.011wt.%N, the austenite grain radius showed little variation over the temperature range 1100°C to 1300°C, the grain radius only varying from 27μm up to 33μm. Whilst this does not show absolute agreement with the calculated values in Table 6.4, only small adjustments of Z and r_0 would eliminate the difference.

Titanium nitride technology has been applied to inhibit grain growth in the coarse grained heat affected zone of fusion welds. Typical results of grain refining additives are shown in Figure 6.15 (Kanazawa, 1976). Titanium nitride is by far the most potent inhibitor of grain growth at temperatures up to the solidus, near to the fusion line of the weld. Similar results have been reported by Roberts (1984), and dramatic improvements in the prior austenite grain size and toughness properties of the HAZ have been produced (Hulka and Heisterkamp, 1977), by adding 0.2wt.% Titanium to a carbon manganese steel as shown in Figure 6.16.

A second application of this technology is found in medium-carbon vanadium steels for the manufacture of forged engineering steel components such as con-rods and steering links in automobiles. The stock material is heated to temperatures of about 1200°C and then hot forged. These steels were developed to give the necessary strength and toughness in the as-forged condition thereby minimising the manufacturing costs. The small sub-stoichiometric addition of titanium proved to be important in avoiding the growth of very large austenite grains on those parts of the forging that received little or no hot deformation. In the absence of hot deformation, the grain size of conventional material is that which exists at 1200°C, i.e. a very coarse grained structure that results in the formation of a coarse brittle microstructure on cooling. The application of titanium nitride technology not only prevents grain growth in these regions, but also inhibits grain growth in the hot deformed regions after recrystallisation.

6.4 HOT-ROLLED PRODUCTS

The application of grain refinement principles to the hot working process, already referred to in the case of titanium nitride technology as applied to medium carbon steel forgings, was based on a considerable body of information deriving from hot rolled structural steels. There are three principal areas where grain refinement practices are relevant.

(i) Controlling the austenite grain size prior to the onset of deformation.

(ii) Inhibiting grain growth after recrystallisation between the various deformations (rolling passes).

(iii) Inhibiting the recrystallisation process itself.

Examples of the application of grain size control will be considered in each of these areas.

6.4.1 Control of the Starting Grain Size

Considerable efforts have been made to develop constitutive equations for the broad evolution of microstructure during multi-stage hot rolling processes, (as discussed in Chapter 4). These enable a prediction of the final microstructures, which are so important to the material properties. They also allow the prediction of rolling mill loads that arise

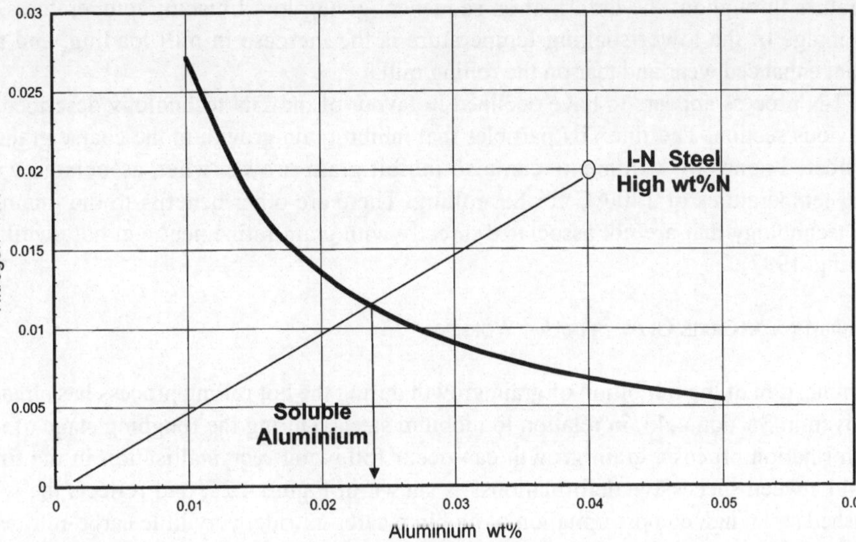

Fig 6.17 Solubility of Aluminium Nitride at 1200°C.

from the metallurgical condition of the steel (progress of recrystallisation) as well as the temperature of deformation. The literature on this subject is extensive, but an extensive overview (Sellars, 1980) summarised much of this constitutive work. Of particular interest here is the equation relating to the recrystallised grain size, d_{rex}, –

$$d_{rex} = 0.5 d_o^{2/3} \varepsilon^{-1} \quad \text{(equ. 4.52)}$$

where d_0 is the prior grain size, and ε is the true strain applied during deformation. In multi-stage deformation, there will naturally be a sequence of d_0 values following each deformation, recrystallisation and grain growth cycle, but for any given sequence of deformations, a finer starting grain size will result in a finer recrystallised grain size.

One of the early attempts to exploit this effect was in the I-N process, in which aluminium nitride was the grain refining agent. In Section 6.2, it was shown that grain coarsening temperatures of about 1050 – 1100°C were possible by increasing the aluminium content up to about 0.05wt.%, and that further additions of aluminium did not give any further benefit in steels containing about 0.01wt.%N. In the I-N process nitrogen contents were increased into the range 0.015 – 0.02wt.%. As can be seen from Figure 6.17, there will be a considerable undissolved aluminium nitride content at 1200°C. Calculation of the aluminium nitride content for a 0.04wt.%Al 0.02wt.%N steel gives a value of f_v = 6.4 x 10^{-4}, and, together with a particle size of 30nm (see Figure 6.6), this would be capable of pinning an austenite grain size of R_o = 9μm. It should be noted that the composition chosen here would give a relatively low dissolved aluminium level compared with the higher aluminium steels with 0.01wt.%N, and would not show the accelerated Ostwald ripening depicted in Figure 6.6. The use of high nitrogen contents to stabilise the aluminium nitride and minimise the Ostwald ripening was supplemented by restricting the soaking temperatures in order to enhance the volume fraction of aluminium nitride and reduce the rate of particle coarsening. The restricted soaking temperatures also lowered the temperature throughout the deformation sequence, giving low finishing temperatures. A disadvantage of the lower soaking temperature is the increase in mill loading, and the attendant enhanced wear and tear on the rolling mill.

The I-N process appears to have declined in favour of the TiN technology described in the previous section. The fine TiN particles that inhibit grain growth in the coarse grained heat affected zones of weldments can also inhibit grain growth when associated with soaking temperatures of 1200°C for hot rolling. There are other benefits to the titanium nitride technology that are not associated directly with grain refinement e.g. hot ductility, (Pickering, 1997).

6.4.2 Inhibition of Grain Growth in Hot Working

Some indication of the inhibition of grain growth during the hot rolling process has already been given in Section 4.43, in relation to niobium steels. During the roughing stage of the hot deformation process, grain growth can occur following recrystallisation in the time interval between successive deformations, as shown in Figure 4.25, and reflects the well established strain induced precipitation of niobium carbo-nitride; very little carbo-nitride is required when the matrix grain size, R_0, is relatively large. More will be said of the effects of niobium in Section 6.4.3.

A particular example of the application of grain growth inhibition was found in connection with rod or bar rolling in a continuous mill. The continuous bar mill is operated a high speeds and provides no opportunity to stop the rolling process in order to cool the stock to a suitable finishing temperature. In fact, in a modern bar mill, particularly in the finishing train, the temperature of the stock is often increasing rather than falling, as there is no time for the dissipation of the heat that is generated by the work of deformation. Finishing temperatures in these mills are commonly in the range of 1000 – 1050°C. The refinement of the grain structure by the hot working process itself, resulting from the rapid deformation and recrystallisation sequences, needed to be matched by a grain refining additive that would inhibit grain growth at the end of the sequence, when the recrystallised grain radius is of the order of $R_0 = 10\mu m$. Titanium nitride technology provided an attractive solution in the case of high strength long products. The strengths were developed by the precipitation of vanadium carbo-nitride during transformation, and the magnitude of this effect is very dependent upon the amount of nitrogen available. The use of 0.01wt.%Ti in steels containing about 0.015wt.%N, would clearly leave sufficient nitrogen for the vanadium strengthening process. This process is also known as recrystallisation controlled rolling.

6.4.3 Retarded Recrystallisation

The introduction of microalloying elements in the late 1950's, and the stimulation of interest in strong tough steels in the as-rolled condition, led to extensive studies of controlled rolling as described in Section 4.4. The effects of microalloys on the recrystallisation temperature (Figure 4.24), and on the development of thin elongated grain structures during conventional low temperature controlled rolling, particularly in niobium steels (Figure 4.28), were noted. Whilst some of the retardation effect by niobium was attributed to the niobium in solution, a major retardation was seen following the strain induced precipitation of niobium carbide. It is interesting to consider the recrystallisation process as one of sub-grain growth. Major differences between sub-grains and grains are their sizes, and the misorientations between neighbours. Sub grains can vary in size but are often in the 0.5 – 1.0μm (diameter) range following substantial deformation. Although normal high angle grains have been observed with diameters down to 5μm, the more commonly observed values (in grain refined austenite for example), are of the order of 10μm – 20μm. The grain misorientations in high angle boundaries, although variable, seem to play no part in the grain growth process of bulk materials. Sub-grains can show different misorientations, but the sub-grain boundary energies are generally much lower, as described in Chapter 1, and variations would also be expected to play little part in the sub-grain growth process, although the remnants of former high angle boundaries in a deformed structure may provide a high boundary energy compared with the sub-grain boundaries (Humphreys, 1999). Nevertheless, it is interesting to establish the characteristics of a particulate array i.e. volume fraction and particle size, required to pin such a fine array of sub-grains.

Consider the case of randomly distributed particles, Equation 4.11.

$$r = R_0 f/(1 - 4/3Z)$$

which, in the case of z = 5/3 becomes –

$$f = r/5R_0 \qquad \text{(equ. 6.6)}$$

If we assume that the finest possible particle radius is about 2nm and the sub grain diameter is 1μm (1000nm), then the volume fraction of particles required is –

$$f = 2/5 \times 500$$
$$= 0.0008 \text{ or } 0.08\% \text{ by volume.}$$

This would equate with a niobium content of 0.068wt.%Nb that would need to be in solution at the soaking temperature, and would need to have precipitated at the deformation temperatures. Whilst this is a possibility with a high solution treatment temperature and deformation below 900°C, it would not explain the results of Jonas and Weiss (1979), in Figure 4.23, at temperatures between 900°C and 950°C, where recrystallisation commences when there is only partial precipitation.

It we consider that the precipitate is strain induced and that particles are formed on the sub-grain boundaries, the appropriate relationship is given by Equation 4.23.

$$f = r^2/(3 - 4/Z)R_0^2$$

Again with Z = 5/3 and with the same values of r and R_0, the volume fraction of particles required is –

$$f = 4 \times 3/5 \times 500 \times 500$$
$$= 0.0000096 \text{ or } 0.001\% \text{ by volume.}$$

This would easily be achieved with only a small fraction of the maximum possible precipitation. However, the most convincing evidence for sub-grain boundary precipitation is the direct observation as shown in Figure 5.15. The particles are about 5nm in diameter, and are arrayed as a network of about 1μm diameter, which presumably reflect the sub-grain structure in a hot deformed austenite grain prior to the allotropic change during cooling. The fact that all of the particles in Figure 5.15 are illuminated by the same NbC reflection suggests that they precipitated with the same orientation relationship within a single austenite grain.

Similar effects of fine particles have been observed in retarding recrystallisation in cold worked and annealed steels and in non-ferrous alloys, (Gladman, McIvor and Pickering, 1971, Mould and Cotterill, 1967).

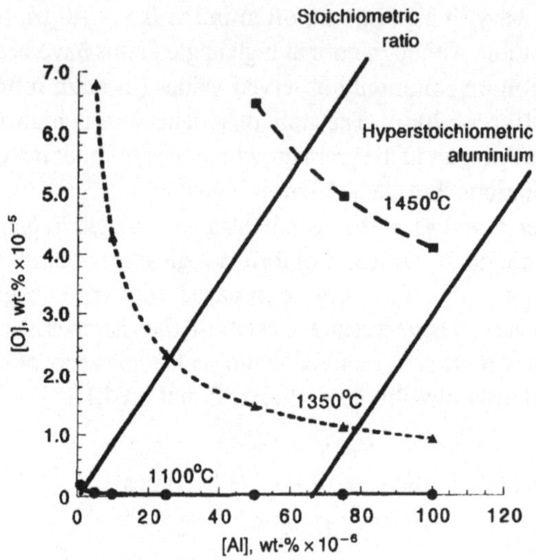

Fig. 6.18 Solubility of Alumina in Austenite.

6.5 SECONDARY RECRYSTALLISATION OF COLD WORKED AND ANNEALED STRIP

The intricacies of cold rolled and annealed aluminium-treated low-carbon steel strip in both the batch annealed and the continuously annealed conditions have been well documented, (Hudd, 1992). Of particular interest were the conditions for the onset of secondary recrystallisation in steels containing aluminium nitride (Dasarathy and Hudd, 1967). Aluminium nitride particles had formed on the grain boundaries of the primary recrystallised ferrite grains during box annealing after cold working, and outlined a grain size of R_o = 5μm. The volume fraction of aluminium nitride, calculated from the nitrogen content was 0.00026. At the start of secondary recrystallisation, the particle size was found to be about 0.10μm. The value of Z calculated on the basis of Equation 4.23 was 1.7. This is very similar to other values of Z reported for AlN and NbC in austenite (Gladman and

Table 6.5 Particle Contributions to Grain Growth Inhibition in a Titanium Deoxidised Steel, (1200°C).

Particle Type	f_i x10³	r_i μm	f_i/r_i μm⁻¹x10³
Ti_2O_3	1.19	3.10	0.38
MnS (fine)	0.115	0.23	0.50
MnS (coarse)	0.35	2.00	0.18
TiN (fine)	0.035	0.016	2.19
TiN (coarse)	0.057	1.00	0.057
All	$10^3\Sigma f_i$ = 1.75	—	$10^3\Sigma f_i/r_i$ = 3.31

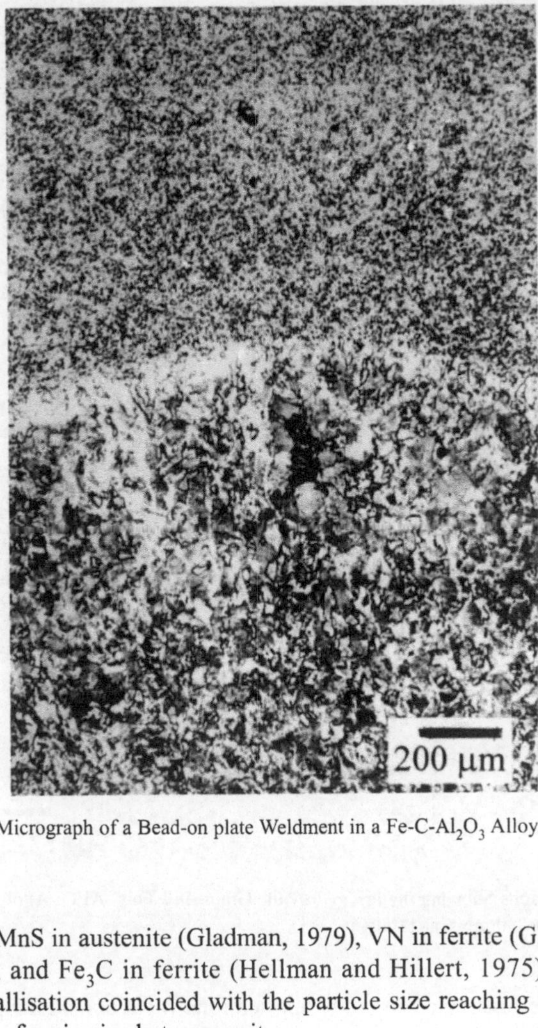

Fig. 6.19 Composite Micrograph of a Bead-on plate Weldment in a Fe-C-Al_2O_3 Alloy.

Pickering, 1967), MnS in austenite (Gladman, 1979), VN in ferrite (Gladman, McIvor, and Pickering, 1971), and Fe_3C in ferrite (Hellman and Hillert, 1975). Thus the onset of secondary recrystallisation coincided with the particle size reaching the critical value for the expected level of grain size heterogeneity.

6.6 THE USE OF OXIDE PARTICLES

The main requirements for grain growth inhibition by second phase particles are –

(i) A very limited solubility of the second phase particles in the matrix at elevated temperatures, so that an adequate fraction of particles is maintained.

(ii) The particles should exhibit very low rates of coarsening by the Ostwald ripening process. The main factors controlling the rate of Ostwald ripening are the diffusivity of the rate limiting specie from one particle to another, and the concentration of this specie in the matrix. Low values of diffusivity and solute concentration are conducive to low coarsening rates.

Oxide particles combine a high thermodynamic stability with extremely slow growth

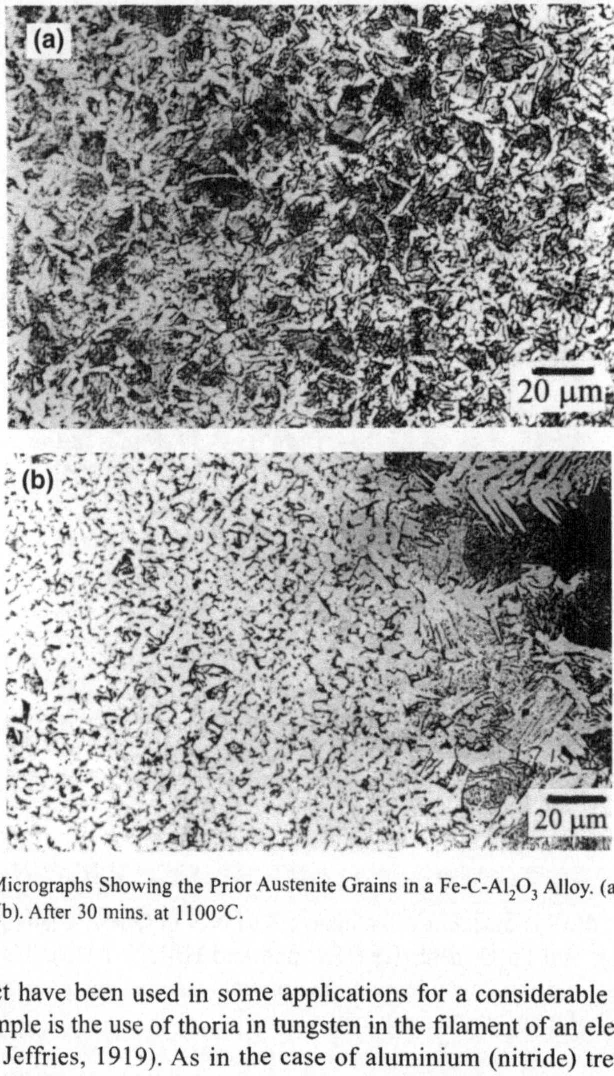

Fig. 6.20 Optical Micrographs Showing the Prior Austenite Grains in a Fe-C-Al_2O_3 Alloy. (a). Near fusion zone of weldment. (b). After 30 mins. at 1100°C.

rates, and in fact have been used in some applications for a considerable period of time. The classic example is the use of thoria in tungsten in the filament of an electric light bulb, (Jeffries, 1917, Jeffries, 1919). As in the case of aluminium (nitride) treated steels, the practice of grain refinement in tungsten preceded any quantitative understanding of grain growth. One of the main problems of the thermodynamically stable oxides in steel is that their very stability promotes their formation in the liquid phase, resulting in the formation of large oxidic particles. The size of these particles is inappropriate for the inhibition of grain growth in fine grained austenite, as was the case for large titanium nitride cuboids at high titanium contents. The size of these oxides is usually of the order of 1 – 10μm, and at volume fractions below 2×10^{-3}, they are only capable of inhibiting growth when $R_o = 100 - 1000$μm, i.e. typically those grain sizes seen at temperatures well above the grain coarsening temperature, Figure 6.1. Detailed studies of the effects of titanium oxides and manganese titanates, (Senogles and Gladman, 1997), showed grain sizes typically in the region of 200μm, and the stability of even this coarse grain size was mainly due to the incidental presence of titanium nitride. This work was particularly interesting in that due account was given to the simultaneous presence of coarse titanium oxides, fine and coarse

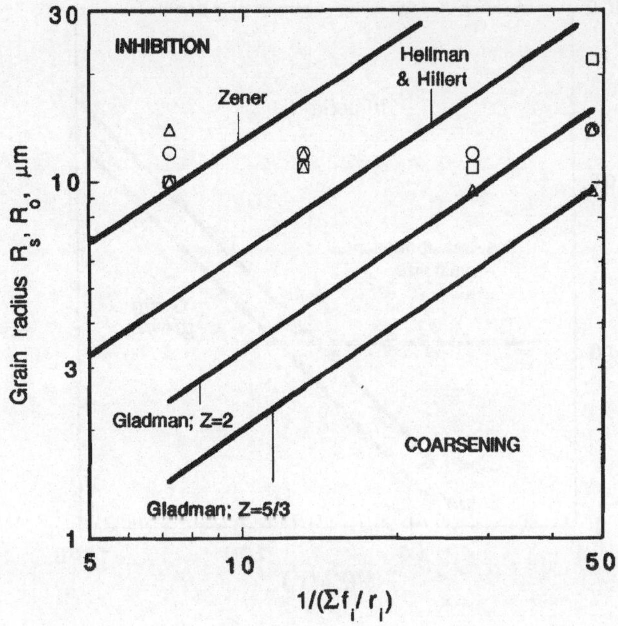

Fig. 6.21 Austenite Grain Sizes in Alumina Doped Steels Compared with Theoretical Models.

manganese sulphides, and fine and coarse titanium nitrides, and their individual contributions to the $\Sigma(f_i/r_i)$ function of Equation 4.29. These contributions are shown in Table 6.5 for an air-melted 0.023wt.%Ti steel held at 1200°C. The dominance of the fine titanium nitride particles is clear, both at this temperature and also at much higher temperatures (up to 1350°C). At 1200°C, the observed grain size of ASTM2, is equivalent to a grain radius of 88μm. The predicted radii with Z values of 5/3 and 2 are 60 and 101μm respectively, which embrace the experimental value.

A further feature of this work was the observation of mixed grain sizes with some abnormally large grains at the higher austenitising temperatures used 1300 – 1350°C. It is quite possible that this is due to the coarsening of the finer TiN and MnS particles, but, because normal grain growth has occurred to establish the grain radius of about 100μm, it is also possible that the arrest of normal grain growth will start with those grains having a marginal advantage until very few grains, i.e. those with the largest size advantage, are still growing. This possibility obviously requires further investigation.

Studies of the effects of spherical silica particles on austenite grain growth confirmed that the large particles of silica formed in the melt would only inhibit austenite grain growth when the austenite grain radius was in excess of 100μm (Weeks, 1997). The main conclusion from this work, and from the earlier work on titanium oxide, was that the virtual insolubility of these oxides in the solid austenite is accompanied by a very limited solubility in the liquid state, and therefore very coarse particles are formed in the liquid, with the resultant coarsening of the austenite grains during subsequent re-heating. Very coarse alumina dendrites have also been observed to form in the liquid state prior to solidification, in steels treated with aluminium.

An alternative method of manufacture, i.e. mechanical alloying, has been used

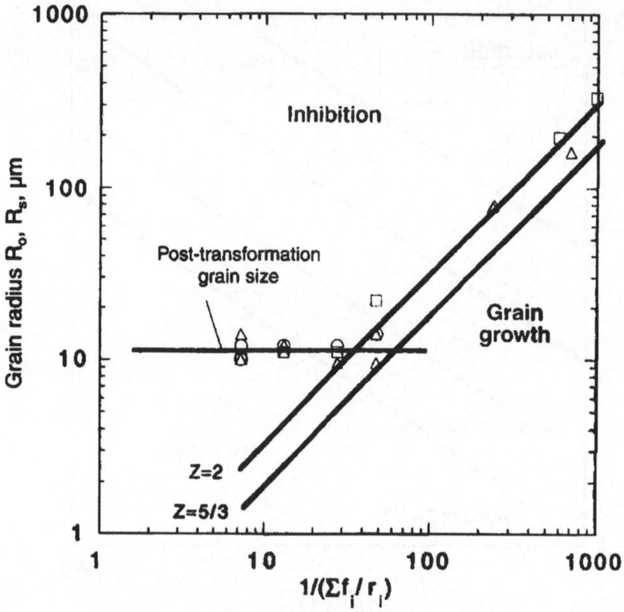

Fig. 6.22 Effects of Particle Size and Volume Fraction of Alumina on Austenite Grain Size Including both Cast and Mechanically Alloyed Steels.

(experimentally) to provide the necessary fine oxide particles in a solid steel matrix (Gladman, Fourlaris, and Talafi-Noghani, 1999). Alumina was selected because of its exceptionally high thermodynamic stability. Published thermodynamic data (Gatellier and Ollette, 1979; Turkdogan, 1989; Swisher and Turkdogan, 1967) was used (Rose, 1995) to obtain the solubility product, k_s, for alumina in austenite, where –

$$k_s = [Al]^2 [0]^3 \quad \text{(equ. 6.7)}$$

$$\log k_s = 9.98 - 53{,}702/T \quad \text{(equ. 6.8)}$$

Even at the solidus temperature of low carbon steels (1450°C) the solubility product of alumina is only 6.5×10^{-22}. Some indication of the low levels of solute content in the solid

Table 6.6 Driving Forces and Pinning Forces.

Nature of Force	Magnitude (Nm^{-2} or Jm^{-3})
	DRIVING
Grain Growth (Surface Energy)	$10^3 - 10^6$
Recrystallisation	$10^5 - 10^7$
Transformation (Chemical)	$10^7 - 10^1$
	PINNING
20 vol % of 10nm particles	2×10^7
0.1 vol % of 10nm particles	10^5
0.1 vol % of 1 μm particles	10^3

state is given in Figure 6.18. In stoichiometric alloys, the soluble aluminium level amounts to less then 1ppm by weight at the solidus temperature. The atom fraction of aluminium is only 1.1 x 10^{-6}. In terms of Ostwald ripening, any change in the particle radius within meaningful times at temperatures in the austenite range is negligible.

Alloys were prepared by milling either iron, carbon, and alumina powders, or iron, carbon, and aluminium powders. In the iron-carbon-aluminium compacts, the aluminium was found to have oxidised during hot compaction, so that all alloys were stoichiometric with regard to aluminium and oxygen contents. Alumina contents were in the range f_v = 0.001 to 0.004. Thermal exposures of 30minutes at temperatures up to 1350°C, and of 10seconds at 1450°C in the high temperatures heat affected zone of a bead on plate weldment, all showed that a fine grained austenite of ASTM8 or thereabouts had been maintained, irrespective of the thermal exposure. A low magnification micrograph of the heat affected zone under the bead on plate weldment, Figure 6.19, shows a complete absence of any coarse grained region. The heat affected zones are seen in the upper three-quarters of the micrograph, the fusion zone in the bottom quarter being separated by a light etching area resulting from the mushy zone. A higher magnification of the near fusion zone structure, Figure 6.20(b), with the fusion zone on the right, confirms the absence of coarse grained regions in the HAZ. The microstructure of a sample of the high alumina steel, quenched from 1100°C, showed austenite grains of about 20µm diameter, outlined by ferrite formed during cooling, Figure 6.20(a).

The outcome of this work is depicted in Figure 6.21 where the observed austenite grain radii, R_0, are compared with the various theoretical predictions (Zener, 1949; Hellman and Hillert, 1975; Gladman and Senogles, 1997). Zener's equation (Equation 4.9) predicts that only the higher alumina alloy should be stable, Hellman and Hillerts prediction is that only two of the four alloys should have stable grain sizes, whilst Equation 4.29 suggests that all four alloys would have stable grain sizes, as indeed they had.

The invariance of the grain sizes shown in Figure 6.23 arises from the high driving force for microstructural change when heating a ferrite-carbide aggregate into the austenite range. The driving forces for microstructural change are summarised in Table 6.6 (Ashby, 1980). Only the highest pinning forces (involving some 20% by volume of fine particles) would retard some of the lower chemical driving forces. Consequently, with the pinning forces used in grain refined steels, 10^3 - 10^5 Jm^{-3}, the particles would have very little effect on a transformation boundary with a driving force of some 10^7 – 10^{11} Jm^{-3}. Thus the 'as-transformed' grain size would be expected to be independent of the presence of particles, but, upon completion of the transformation, grain growth would be inhibited, depending upon the precise pinning forces and driving forces as expressed by Equations 4.11 or 4.29 (or their counterparts for grain boundary precipitates).

A broader overview of the effects of oxides (and indeed all second phase particles) is given in Figure 6.22, which includes the results obtained from the mechanically alloyed Fe – Al_2O_3 materials, and the conventionally manufactured Fe – SiO_2 and Fe – Ti_2O_3 alloys. If we accept that the allotropic transformation is unaffected by the fractions and sizes of particles present in these steels, and that the austenite grain size so formed is of the order of ASTM8, then the stability or otherwise of this grain size will be dictated by the pinning action of the particles. High pinning forces will inhibit grain growth, whilst low pinning forces allow grain growth to occur. The borderline between these zones is given, for randomly distributed particles, by Equations 4.11 (for uniformly sized particles) or 4.29 (for distributed particle sizes). The borderlines for likely values of grain size heterogeneity of Z = 5/3 and Z = 2 are shown in Figure

6.24. Grain sizes above these lines will result in inhibition of growth, whilst those initially below the line will undergo growth until they reach the borderline at a level of grain size dictated by the pinning particles. If we assume that all of the steels gave rise to an 'as-transformed' austenite grain size of about 10μm, then those steels with particle fractions and sizes having values of $1/\Sigma(f_i/r_i)$ less than about 50 would have stable grain sizes, whilst those giving values greater than 50 undergo grain growth. The value of $1/\Sigma(f_i/r_i)$ below which the transformed grains will be stable is clearly dependent upon their size. A transformed grain size of only 3μm for example would require a $1/\Sigma(f_i/r_i)$ value of less than 10 to inhibit grain growth.

Figure 6.22 should be applicable to all systems with small volume fractions of randomly distributed particles, and a similar diagram could be constructed for grain boundary precipitates. The only exception in terms of inhibition/growth would be in the case of abnormal grains, where of course the values of Z would be considerably higher than those used in the construction of Figure 6.22.

REFERENCES

Adrian H., 1995. Microalloying '95, Pittsburgh, June, p.285.

Bain E.C. and Paxton H., 1961. Functions of Alloying Elements in Steels. ASM, Metals Park, Ohio.

Chatterjee A.B. and Nijhawan B.R., 1957. Metal Treatment, vol.24, p.54.

Dasarathy C. and Hudd R.C., 1967. Acta Met, vol.15 p.1665.

George T. and Irani J.J., 1968. Journ. Australian Inst. Metals, vol.13, p.94.

Gladman T., 1963. Developments in Carbon Steels, Iron Steel Inst., Special Report No.81, London, p.68.

Gladman T. and Pickering F. B., 1967. Journ. Iron Steel Inst., vol.205, p.653.

Gladman T., McIvor I.D. and Pickering F. B., 1971. Journ Iron Steel Inst., vol.209, p.380.

Gladman T., 1979. Inst. of Metallurgists, Monograph No.3 (inclusions). Ed. F. B. Pickering, City Polytechnic, Sheffield.

Gladman T., 1992. Eds. G. Tither and S. Zhang, TMS, Warrendale, PA. P.3.

Gladman T. and Senogles D.J., 1997. Titanium Technology in Microalloyed Steels. Ed. T.N. Baker, Inst. of Materials, London, p.83.

Gladman T., Fourlaris G. and Talafi-Noughni M., 1999. Materials Science and Technology, vol.15, p.1414.

Hall E.O., 1951. Proc. Phys. Soc., vol.64B, p.747.

Hellman P. and Hillert M., 1975. Scandinavian Journ. Met., vol.4, p.211.

Honer K.E. and Baliktay S., 1977. 44th Int. Foundry Conf., Florence, paper 11.

Hudd R. C., 1992. Materials Science and Technology, vol. 7, (Constitution and Properties of Steel). Ed. F. B. Pickering, VCH, Weinheim, p.219.

Hulka K. and Heisterkamp F., 1997. Titanium Technology in Microalloyed Steels. Ed. T. M. Baker, Institute of Materials, London, p.65.

Humphreys F. J., 1999. Materials Science and Technology, vol.15, p.37.

Irvine K.J., Pickering F.B. and Gladman T., 1967. Journ. Iron Steel Inst., vol.205, p.563.

Jeffries Z., 1917. Trans. Amer. Inst. Met. Eng., vol.56, p.571.

Jeffries Z., 1919. Trans. Amer. Inst. Met. Eng., vol.60, p.588.

Kanazawa S., 1976. Trans. Iron and Steel Inst., Japan, vol.16, p.486.

Konig P., Schulz W., and Ulmer H., 1961. Archiv. f.d., Eisenhuttenwes, vol.32, p.541.

Leslie W. C., Rickett R. L., Dotson C. L. and Walton C. S., 1954. Trans. ASM. vol.46, p.1470.

Matsuda S. and Okamura N., 1978. Trans. Iron and Steel Inst., Japan, vol.18, p.198.

Mould P. and Cotterill P., 1967. Journ. Mat. Sci., vol.2, p.241.

Narita K., 1975. Trans. Iron and Steel Inst., Japan, vol. 15, p.145.

Osuzu H., 1986. SAE, International Congress and Exhibition, Technical Paper 860131.

Petch N.J., 1953. Journ. Iron and Steel Inst., vol.174, p.25.

Pickering F. B., 1997. Titanium Technology in Microalloyed Steels. Ed. T. N. Baker, Institute of Materials, London, p.10.

Priestner R., Zhou C. and Ibraheem A.K., ibid. p.150.

Roberts W., 1984. HSLA Steels-Technology and Applications. Ed. M. Korchynsky, ASM, Metals Park, Ohio, p.33.

Sellars C.M., 1980. Hot Working and Forming Processes. Eds. C.M. Sellars and G. J. Davies, The Metals Society, London, p.3.

Weeks R., 1997. Grain Control in Steels by Silica Particles, Ph.D. Thesis, University of Leeds.

Answers to Exercises

Chapter 2

1 (a) The Schlegel diagram for a cube is –

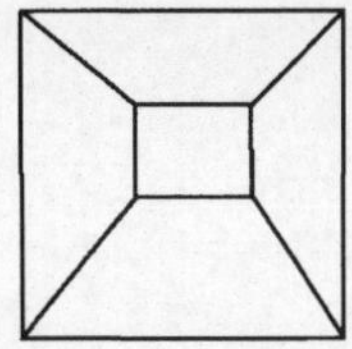

1 (b) The Schlegel diagram for an octahedron is –

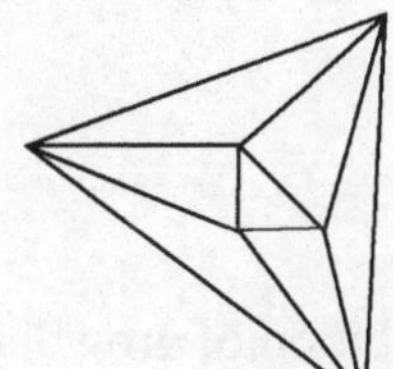

2. From the mathematical expression for the ASTM grain size (Equation 2.38)

$$n = 2^{G-1}$$

where n is the number of grains per square inch at a magnification of x 100 diameters

$$n = 2^5 = 32$$

and therefore the number of grains per square millimetre, n_a, is given by –

$$n_a = (32 \times 100 \times 100)/(25.4 \times 25.4)$$
$$= 496 \text{ mm}^{-2}$$

The mean area, a, is given by –

$$a = 1/n_a = 0.002 \text{ mm}^2$$

The Jeffries number is defined as the number of grains per square millimetre, i.e. n_a, so that

$$n_a = 496\text{mm}^{-2} = \text{the Jeffries number.}$$

3. (a) The standard deviation for a binomial distribution is given by Equation 2.19 as

$$\sigma = [f(1\text{-}f)/n]^{1/2}$$

If f is 0.5 and n is 400

$$\sigma = [0.5 \times 0.5/400]^{1/2}$$
$$= 0.025$$

(The 95% confidence limits are approximately 0.5 + 2s and 0.5 – 2s, i.e. 0.45 and 0.55).

3. (b) The standard deviation for lineal analysis is given (Equation 2.21) as

$$\sigma = f_\alpha(1-f_\alpha)(2/n_\alpha)^{1/2}$$

$$\text{for } \sigma = 0.025 \text{ and } f_\alpha = 0.5$$

$$(n_\alpha/2)^{1/2} = 0.5(1-0.5)/0.025$$

$$= 10$$

Therefore - $n_\alpha = 200$

Thus the number of α particles to be intercepted to give a standard deviation of 0.025 is 200.

Chapter 4

1. For the inhibition of growth by randomly distributed particles of uniform size, Equation 4.11(b) is appropriate.

$$f_v = (1-4/3Z)\ r/R_0$$

When r = 10nm, R_0 = 10μm = 100000nm, and Z = 5/3, then –

$$f_v = 10/(10000 \times 5)$$

$$= 0.0002$$

A volume fraction of 0.0002 or more will be sufficient to inhibit growth of the 10μm radius grains.

2. For grain boundary distributed particles Equation 4.23 is appropriate.

$$f_v = (3-4/Z)\ r^2/R_0^{\ 2}$$

$$= 0.2 \times 10^{-6}$$

3. To find the effect of changing Z from 5/3 to 2 again using Equation 5.11(b)

$$f_v = 10/(10000 \times 3)$$

$$= 0.33 \times 10^{-3}$$

(Compare this answer with that for question 1. The increase in Z has resulted in a 67% increase in the volume fraction of particles required).

4. When there is more than one particle size involved, Equation 4.29 can be used.

$$R_0 = (1-4/3Z)/\Sigma(f_i/r_i)$$

Now

$$\Sigma(f_i/r_i) = (2 \times 10^{-4}/10) + (2 \times 10^{-4}/20)$$

$$= 0.3 \times 10^{-4}\ \text{nm}^{-1}$$

Therefore, with Z = 5/3 –

$$R_0 = 1/(5 \times 0.3 \times 10^{-4})$$

$$= 6{,}667\mu m$$

The particles will stabilize grains with a 6.7μm radius or more.

(Compare this with the 10μm in question 1, having doubled the particle fraction.)

Chapter 5

1. (a) The equilibrium constant is obtained from

$$\log k_s = A - B/T$$

which for vanadium nitride (Table 5.2) is

$$\log k_s = 3.02 - 7840/T$$

when T is 900°C, or 1173°K

$$\log k_s = 3.03 - 7840/1173$$
$$= -3.664$$

Therefore $$k_s = 2.17 \times 10^{-4}$$

(b) The data must be substituted into a form of equation 5.23 to obtain the quadratic with

$$a = 14/51$$
$$b = -(14V_T/51 + N_r)$$
$$c = N_T V_T - k_s$$

Solution of the quadratic gives –

$$V_{VN} = 0.026\text{wt.\%}$$

(c) $$\text{\% by wt. of VN} = V_{NV} \times (51 + 14)/51$$
$$= 0.033\text{wt.\%}$$

(d) $$f_{VN} = 0.033 \times \rho_{Fe}/100\rho_{VN}$$

(Note: This is the approximate solution) The densities can be obtained from Table 5.3 or 5.8 to give –

$$f_{VN} = 0.00044$$

2. This problem clearly involves the quartic equation (5.39) and is solved by iteratively replacing values of [Nb] i.e. the dissolved niobium content. A computer program can be devised for automatic solution, or alternatively, the Microsoft XL spread sheet can be formulated for a manual iteration to give a graph similar to Fig. 5.10, but for 1000°C. The dissolved niobium level is approximately 0.011wt.%, so that the niobium as Nb(C,N) is therefore 0.03 – 0.011 or 0.019wt.%

   The carbon and nitrogen in solution can be obtained from the soluble niobium content and the solubility products.

$$[C] = k_C/[Nb]$$
$$[N] = k_N/[Nb]$$

The carbon and nitrogen as carbo-nitride are obtained from the mass balances –

$$C \text{ as } Nb(CN) = C_T - [C]$$
$$N \text{ as } Nb(CN) = N_T - [N]$$

– and the carbide fraction (on a weight basis) is the quotient, which for the question given, is about 0.66.

3. (a) Using Narita's data

$$k_s = 3.82 - 15.020/T$$

At 1250°C (1523°K) $\quad k_s = 9 \times 10^{-7}$

Ti as TiN is obtained from the quadratic equation (5.23) with

$$a = 14/48$$
$$b = -(14Ti_T/48 + N_T)$$
$$c = N_T Ti_T - k_s$$

This gives a value for the titatium as TiN of 0.0097wt.%. Thus –

[Ti] is $\quad [Ti] = 0.01 - 0.0097\text{wt.\%}$
$= 0.0003\text{wt.\%}$

(Due to the very small value, errors can easily be introduced by the slightest approximations).

3. (b) Coarsening of the 10nm particles is given by Equation 5.54, as

$$r^3 = r_o^3 + 8D\sigma\Omega St/9RT$$

S is expressed as an atom fraction, so –

$$S = 0.00003 \times 56/(48 \times 100)$$
$$= 0.3 \times 10^{-6}$$

D for titanium is obtained from Table 5.5 as –

$$D = 0.15 \times 10^{-4} \exp(-251200/RT)$$
$$= 3.6 \times 10^{-14}$$

Using other constants given in the text and Table 5.6 (note that all the units of length must be in metres and units of time in seconds) –

$$r^3 = r_o^3 + (8)(3.6 \times 10^{-14})(0.8)(11.44 \times 10^{-6})(0.3 \times 10^{-6}) \times 11800/(8.314 \times 1523)$$
$$= 10^{-24} + 73.7 \times 10^{-26} m^3$$

To convert this to nm^3, multiply by 10^{27} to give

$$r = 12\text{nm}$$

Index

A

Abnormal Grain Growth 61, 62, 88, 94, 95, 150
Aboav-Weaire Law 59, 60
Allotropic change 96, 100, 131, 143, 144, 167
Alumina 171, 172, 173
Aluminium Nitride 165, 168
Anisotropy 11, 59
Area fraction 31, 32
Areal analysis 32
Areal intercept 34, 35, 36, 42, 55
Arrhenius relationship 102, 114, 137
ASTM Standard E112 40, 41
Asymmetric tilt boundary 3
Atomic mass 116, 140, 158

B

Binding energies 16
Boundary forces – see grain boundary forces i
Bulk diffusion viii, x, 134, 135, 136, 139, 140, 152, 158

C

Capping 92, 93
Carbo-nitrides 111, 120, 131, 132, 137, 139
Coincident site lattice 4, 5, 7, 8, 18
Comparator chart 40
Continuous cooling transformation 98
Controlled rolling 100, 101, 102, 104, 166
Cube viii, 21, 27, 34, 36, 37, 38, 43, 114, 177
Curvature 5, 49, 56, 59, 60, 61, 88, 133

D

Density viii, 6, 50, 51, 87, 105, 129, 130, 138
Diffusion coefficients 137, 140, 152
Dislocations 2, 3, 4, 15, 17, 33, 47, 102, 134, 136
Displacement Shift Complete Lattice 5
Dodecahedron 26, 27, 28, 34, 36, 38, 39, 59

E

Engineering steels 164
Equilibrium diagram 96
Euler's Law 21, 22

F

Faceted boundaries 10
Forces – see grain boundary forces i
Free surfaces 67, 92

G

Grain boundaries viii, ix, 1, 2, 4, 8, 10, 11, 13, 15, 16, 23, 29, 32, 33, 40, 47, 50, 58, 62, 67, 68, 77, 78, 81, 84, 102, 104, 107, 131, 132, 136, 143, 154, 167, 168
Grain boundary diffusion viii, 134, 136
Grain boundary dislocations 4, 6, 15, 17
Grain boundary energy 6, 7, 8, 12, 14, 15, 17, 18, 23, 47, 48, 53, 56, 58, 67, 68, 75, 85, 87, 91, 94
Grain boundary segregation 15
Grain growth ix, 16, 28, 47, 48, 49, 50, 51, 53, 54, 55, 57, 58, 59, 60, 67, 69, 71, 72, 73, 74, 77, 78, 79, 81, 83, 84, 85, 86, 87, 88, 89, 91, 92, 93, 94, 95, 97, 99, 100, 101, 103, 104, 105, 129, 143, 144, 146, 149, 150, 151, 156, 157, 163, 164, 166, 170, 173
Grain shapes 21, 27, 28, 38
Grain size viii, 29, 30, 40, 43, 61, 62, 67
Grooving – see Thermal grooving i
Growth velocity 62

H

Heat affected zones 162, 166, 173
Heat treatment 151
Heyn intercept 40, 41, 43
Hot rolling 102, 103, 105, 120, 131, 154, 157, 164, 166

I

Icosahedron 38, 39
Impurities viii, 8, 15, 16, 62, 63, 80, 82
Intercept 31, 32, 33, 37, 38, 41, 43, 44, 55, 73, 98, 107, 118, 178
Interfacial area 47, 48, 107, 133
Interfacial tension 10
Isotropy 59

J

Jeffries planimetric method 42

L

Lineal analysis 31, 178
Linear intercept 32, 34, 35, 37, 39, 42, 43, 55, 67
Low angle boundaries 5, 7

M

McQuaid – Ehn grainsize 40
Mismatch 3, 5, 18, 139
Mobility 53, 56, 63
Molar volume 134, 135, 136, 138, 140, 152

N

Niobium carbide 73, 82, 101, 120, 121, 122, 125, 128, 166
Niobium carbo-nitride 101, 124, 128, 129, 130, 132, 133, 157, 166
Niobium nitride 120, 125, 127, 128
Normal grain growth 48, 61, 62, 88, 94, 148, 149, 150, 171
Normalising 96, 97, 98, 100, 105, 143, 157, 158

O

Octahedron 22, 38, 177
Ostwald ripening ix, 80, 111, 131, 133, 134, 146, 147, 153, 156, 158, 159, 160, 162, 163, 165, 169, 173
Oxides 151, 170, 171, 173

P

Particle growth 111, 136, 140, 160
Particle pinning 68
Particle size 71, 72, 74, 77, 78, 81, 82, 83, 111, 122, 131, 132, 135, 137, 145, 147, 152, 153, 155, 156, 157, 162, 165, 168, 172, 173, 178
Pinning 70, 82, 103, 158, 174
Point counting 31, 32
Polyhedra 22, 23, 24, 27, 35, 37, 38, 39, 41, 48, 84
Potts model 58, 76

Q

Quadratic equation 119, 124, 140, 180
Quartic equation 126, 179

R

Recrystallisation ix, 101, 104, 105, 148, 164, 165, 166, 167, 168, 172
Recrystallisation controlled rolling 104, 166
Recrystallisation temperature 104, 105, 166

S

Schlegel diagram 22, 177
Secondary recrystallisation ix, 148, 168
Shear bands 105
Silica 171
Size distribution 55, 59, 60, 73, 80, 94, 149, 155
Solubility 16, 17, 18, 80, 82, 96, 112, 113, 114, 115, 116, 117, 119, 120, 121, 122, 124, 127, 128, 136, 139, 140, 147, 152, 158, 159, 160, 163, 165, 168, 169, 171
Solubility Product 112, 113, 114, 115, 116, 118, 119, 120, 121, 123, 124, 125, 127, 128, 136, 138, 157, 158, 159, 162, 172, 179
Solute drag 62, 84
Space filling 1, 23, 24, 26, 27, 28, 38, 41
Sphere 35, 36, 37, 43, 44, 53, 70, 74, 75, 93, 100
Statistical errors 31, 41, 42
Stereological relationships 30
Stereology 21
Stoichiometry 118, 127, 146
Substrates 92
Surface tension 10, 23, 25, 28, 51
Symmetrical tilt boundaries 6

T

Tetrahedron 27, 37, 38, 52, 53, 54, 55, 74, 75, 76
Tetrakaidecahedron 22, 23, 25, 26, 27, 28, 34, 35, 36, 37, 38, 48, 51, 54, 55, 71, 76, 77, 79, 149
Thermal cycles 84, 96, 100
Thermal grooving 67, 85, 88, 89, 94, 100, 150
Thermodynamics111
Thermo - mechanical treatment 100, 111
Tilt boundaries 6, 7, 8
Titanium Carbide 120, 128
Titanium nitride 82, 112, 120, 121, 122, 128, 140, 157, 158, 160, 162, 163, 164, 170
Titanium nitride technology 122, 158, 163, 164, 166
Titanium oxide 170, 171
Topological growth models 56
Topological relationships 23, 24
Topology 21
Transformation 40, 47, 48, 53, 76, 97, 98, 99, 106, 107, 123, 132, 133, 139, 143, 151, 155, 158, 166, 173
Tungsten 170
Tungsten – thoria 87
Twist boundaries 2, 3, 4, 6, 10, 18

V

Vanadium carbide 120, 121, 122, 128, 139
Vanadium nitride 70, 120, 123, 128, 140, 157, 160, 179
Vertex model 59, 60, 61
Volume fraction ix, 30, 31, 32, 33, 34, 35, 69, 71, 72, 74, 76, 78, 79, 80, 81, 82, 111, 128, 129, 130, 133, 135, 136, 138, 145, 146, 151, 153, 155, 162, 165, 167, 170, 172, 174, 178

W

Wagner equation 135, 137, 139, 147, 152, 156, 158, 162
Welding 164

Z

Zener relationship 69, 71
Zener – Holloman parameter 102